Clinical Thermology

Subseries Thermotherapy

M. Gautherie (Ed.)

Methods of Hyperthermia Control

With Contributions by
T. C. Cetas · T. V. Samulski · P. Fessenden · J. C. Bolomey
M. S. Hawley · M. Chive

Foreword by T. C. Cetas

With 125 Figures and 7 Tables

Springer-Verlag Berlin Heidelberg New York
London Paris Tokyo Hong Kong

Dr. Michel Gautherie
Laboratoire de Thermologie Biomédicale
Université Louis Pasteur
Institut National de la Santé
et de la Recherche Médicale
11, rue Humann
67085 Strasbourg Cedex, France

ISBN-13: 978-3-642-74638-3   e-ISBN-13: 978-3-642-74636-9
DOI: 10.1007/978-3-642-74636-9

Library of Congress Cataloging-in-Publication Data
Methods of hyperthermia control / M. Gautherie (ed.); with contributions by T. C. Cetas ... [et al.];
foreword by T. C. Cetas. p. cm. − (Clinical thermology. Subseries thermotherapy) Includes biblio-
graphical references.

1. Cancer − Thermotherapy − Standards. 2. Body temperature − Regulation. 3. Thermometers and
thermometry, Medical − Standards. I. Gautherie, Michel. II. Cetas, T.C. (Thomas C.) III. Series.
[DNLM: 1. Hyperthermia, Induced − methods. 2. Neoplasms − therapy. QZ 266 M5925]
RC271.T5M48   1990   616.99'40632 − dc20   89-26190

© Springer-Verlag Berlin Heidelberg 1990
Softcover reprint of the hardcover 1st edition 1990

2127/3145-543210 − Printed on acid-free paper

Foreword

The enormous potential that hyperthermia has for benefiting patients with cancer is impressively indicated by biological studies, both in vitro and in vivo, and by comparative clinical studies whenever the heat has been appropriate to the size of the tumor. But hyperthermia, as with any other technologically based medical procedures, requires an extensive development of sophisticated instrumentation and techniques to offer routine clinical benefit. We probably erred in starting clinical trials so soon. We had hoped that by showing the clinical benefits on some superficial tumors quickly, financial support would be stimulated for the required technological developments.

Unfortunately, treating superficial disease adequately was more difficult than we had supposed and regional treatments were less successful than we had wished. The physical reasons are clear and were apparent from the beginning, although in our enthusiasm we ignored them.

Circumstances are different now. We have to treat a wide range of tumors in various sites, but the systems and techniques required are only available in a few laboratories and clinics where they still are undergoing refinement. Thermal dosimetry, as currently practiced in the best centers (Fessenden and Samulski, Part I) is difficult, tedious, and traumatic. In the future, it will be less so (Bolomey and Hawley, Part II) as heating systems are integrated with noninvasive dosimetry (SAR and thermometry). An example of such an approach is given by Chive (Part III), who shows that in clinical practice reduced trauma is possible for superficial tumors. Nevertheless, in research centers striving to address the dose-response relationship in a detailed fashion, the tedious and traumatic procedures outlined in Part I are still required, even for superficial sites.

To quickly review this volume, in Part I "Thermometry in Therapeutic Hyperthermia," Samulski and Fessenden review the status in the clinical setting. An emphasis is placed on techniques for locating thermometer probes and is followed by a review of various types of thermometers. It can be concluded that while further refinements are desirable, the current status of probe thermometry, both instrumentation and practice, is adequate. Quality control is the greatest problem. However, locating the probes in three dimensional coordinate space, along with the heating applicators and target volumes, is excessively laborious. New procedures and probably new devices must be developed so that this "geometrical dosimetry" can become routine for every patient everywhere.

In Part II, "Noninvasive Control of Hyperthermia," Bolomey and Hawley picked up the gauntlet on non-invasive thermometry that I threw down a few years ago. They have reviewed the various approaches by addressing the compromises concerning sensitivities (thermal, spatial, temporal), the problem of noise rejection (i.e., sensitive devices in a noisy environment), the space constraint (heater and imager targeted at the same volume), and the sensitivity of the thermometric parameter to other physiological variables. They show that the first three of these are merely technical, but nontrivial, and that real solutions are possible, indeed probable as computers get bigger, faster and cheaper. They remind those developing the new methods to consider the sensitivity to physiological variations. If hyperthermia stress changes the parameter under investigation, then that parameter is probably not a good thermometer, but, it may be an excellent indicator of the adequacy of the therapy — a blessing in disguise.

Finally, Part III, "Temperature Measurement by Microwave Radiometry in Hyperthermia Process: Application to Thermal Dosimetry," by Chive describes the development and practical implementation of an integrated heating and noninvasive monitoring system for superficial hyperthermia. The significance is two-fold. First, in its own right, it improves the practice of clinical hyperthermia for superficial disease while reducing trauma to patients. Second, and perhaps more importantly it shows that noninvasive dosimetric devices must be integrated with specific heating approaches for specific sites. An all encompassing heating system probably does not exist, and neither does such a non invasive thermometry system. But several possible systems can be developed in which the patient, the heating approach, and the dosimetry system are all integrated into a totally engineered package.

Overall, this is an impressive volume.

Tucson, January 1990   T. C. CETAS

Contents

List of Contributors

J.C. Bolomey
Laboratoire des Signaux et Systèmes, Ecole Supérieure d'Electricité, Plateau du Moulon, 91190 Gif sur Yvette, France

T.C. Cetas
Department of Radiation Oncology, Health Sciences Center, The University of Arizona, Tucson, Arizona 85724, USA

M. Chive
Centre Hyperfréquences et Semi Conducteurs, Université des Sciences et Techniques de Lille I, Bat. P 4, 59655 Villeneuve d'Ascq Cedex, France

P. Fessenden
Stanford University Medical Center, Department of Radiology, Stanford, California 94305, USA

M.S. Hawley
Barnsley District General Hospital, Department of Medical Physics and Clinical Engineering, Gawber Road, Barnsley S75 2EP, England

T.V. Samulski
Duke University Medical Center, Box 3085, Durham, North Carolina 27710, USA

1 Thermometry in Therapeutic Hyperthermia

T. V. SAMULSKI and P. FESSENDEN

1.1 Introduction

The delivery of local and regional therapeutic hyper-thermia in clinical practice proves to be a difficult challenge. The objective is to elevate the temperature of malignant tissue to a uniform level in excess of some minimum therapeutic temperature (e.g., 42.0 °C) for a period on the order of 30–60 min. At the same time, normal tissue temperatures should be maintained at subtherapeutic levels. This objective is seldom achieved. The difficulty is associated not so much with excessive heating of normal tissue (normal tissue toxicity is thus far reportedly low) as with achieving a minimum, uniform therapeutic temperature in the target tissue volume. It is known that thermal cytotoxicity is very temperature dependent [1]. In addition, thermal tolerance, the ability of cells to become resistant to elevated temperature, is another phenomenon that is highly dependent on temperature and exposure time and may be therapeutically relevant [1, 2]. Thus, the failure to deliver hyperthermia therapy at sufficiently high temperature and with good temperature uniformity may be why heat treatments alone are not very clinically effective.

The sensitization of cells to the effects of radiation and drugs may also occur at elevated temperature [1]. When combined with these conventional therapies the current limitations in delivering hyperthermia are perhaps of lesser clinical consequence. This is particularly true when hyperthermia is used as an adjuvant to radiation. In combination, radiation and hyperthermia therapy can play complementary roles. Hyperthermia is more effective in poorly vascularized regions of malignant tissue, where higher temperatures are more easily reached due to the lack of heat dissipation by blood flow. Concurrently, it is these poorly vascularized tumor regions that are likely to be radiation resistant due to hypoxia. Other biological factors such as cell kinetics can also contribute in a complementary or synergistic way toward enhancing the combined modalities. Therefore, although temperature uniformity is desirable, it may not be necessary for hyperthermia to be a clinically effective adjuvant [3].

Despite the fact that adjuvant hyperthermia has been demonstrated to be efficacious in clinical settings [4], many questions remain with regard to treatment quantification and optimization. There appears to be a strong dependence of tumor response on tumor volume: larger tumors are less likely to be controlled [5–9]. This again may be the result of poor hyperthermia delivery. Typically, the variation in parameters of possible prognostic significance (e.g., mean, maximum, minimum temperature), as indicated by measured standard deviations, are on the order of 1°–2 °C [3]. This clearly indicates the need for better technologies for inducing and controlling hyperthermia. In order to make progress in these areas a more complete characterization of the temperature distributions during therapy is required.

At present the only approach available for acquiring necessary temperature data during clinical hyperthermia treatments is the use of invasive probes. For reasons of both biological sensitivity and the variance of achieved temperatures such probes must have temperature accuracy and resolution on the order of one-to two-tenths of a Celsius degree. In addition to this accuracy and resolution, there are other functional requirements. The thermal gradients observed during clinical hyperthermia indicate a need for spatial resolution on the order of a few millimeters. A probe's thermal conductivity should approximate that of tissue in order to prevent thermal shunting or smearing. It is necessary that these probes function in intense ultrasound (US) and/or electromagnetic (EM) fields, since these modalities are used to induce hyperthermia. Other factors such as probe response time, mechanical durability, cost, and convenience for clinical use must also be considered. Table 1.1 (adopted from reference [3]) lists minimum performance goals desirable for clinically acceptable invasive thermometry systems.

A multiplicity of invasive thermometry concepts and techniques have been proposed for specific application in hyperthermia therapy [10–24]. Not all have

Table 1.1. Thermometry performance goals

Parameter	Minimum performance goal
Calibration accuracy	$\leq \pm 0.2\,°C$ over hyperthermic range $(30° - 60°C)$
Resolution	$\leq \pm 0.2\,°C$
Drift	$\leq \pm 0.1°/h$
Recalibration period	$\geq 24\,h$
Response time	$\leq 4\,s$
Bend artifact	$\leq 0.1\,°C$ for 5-mm bend radius
EM and/or US artifact	$\leq 0.1\,°C$
EM interference	$\leq 0.1\,°C$ for $10\,mW/cm^2$ EM exposure
Durability	Suitable for multiple implantations and mechanical mapping
Thermal smearing	Smearing length $\leq 1.5\,mm$ when tested in $10\,°C/cm$ step thermal gradient [62]
Date acquisition I/O	$\leq 10\,s$ update time

been applied in clinical practice and only a few are available commercially. These include thermometry

systems that utilize thermocouples [19, 24, 25], thermistors [11], gallium arsenide (GaAs) optical absorption sensors [13, 22], and photoluminescent-efficiency and decay-time optical sensors [15 – 17, 23]. No single system has been demonstrated to be acceptable for use in both EM and US fields without measurement artifact. This is largely the result of problems associated with US-induced artifact, which have not been resolved for any of the currently used clinical temperature probes. Several systems meet the performance goals for the EM hyperthermia heating techniques. For US heating, small bare metal probes, such as thermocouples embedded in a needle, produce the minimum measurement artifact.

Clinical techniques for the use of invasive thermometry and current approaches to temperature control in therapeutic hyperthermia are presented in this chapter. The currently available technologies for invasive temperature measurement in hyperthermia are reviewed. The basic need for quality thermometry in clinical practice is discussed along with possible causes of erroneous measurements. Finally, future developments in noninvasive thermometry and thermal modeling of in vivo temperature distributions are also outlined. All of these topics have been addressed in varying degrees of detail throughout the literature [25 – 32].

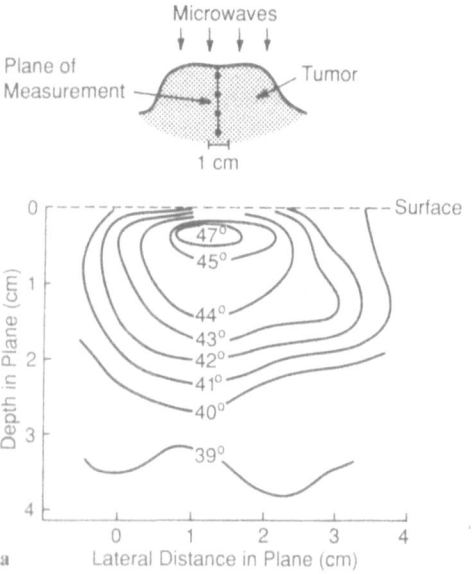

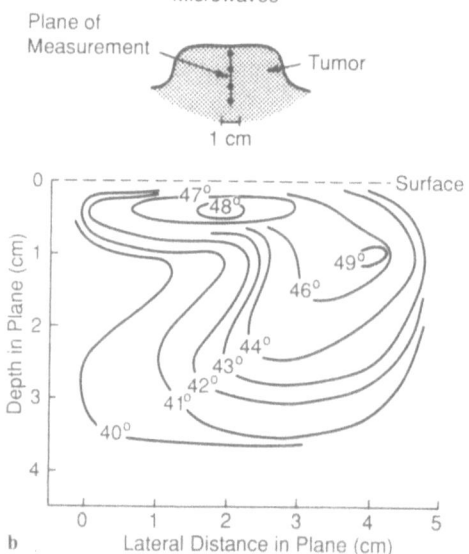

Fig. 1.1a, b. Temperature distributions reconstructed from invasive measurements made during the hyperthermia treatment of a superficial lesion using a MW heating device. The temperature data were acquired by moving (mapping) single-sensor temperature probes through invasively placed catheters (indicated by *solid dots* in the tumor schematic). The catheter array defines a central vertical plane within the tumor volume. The parallel grid of temperature measurement points was used to estimate the two-dimensional distributions for two separate treatments of the same lesion (**a, b**). The distribution in **a** mimics the energy deposition pattern expected from the MW applicator. The distribution in **b** shows an anomalous lateral hot spot

1.2 Clinical Considerations

Techniques and procedures for clinical thermometry are varied, and efforts to standardize even the simplest hyperthermia treatments have only recently been proposed [33, 103, 104]. Current clinical practices range from treatments without any direct temperature measurements to aggressive invasive procedures using multisensor probes and mechanical mapping [3, 22–25, 34, 35]. Invasive measurements have obvious limitations; furthermore, there is a predisposition toward making these invasive measurements in tumor rather than normal tissue, since documentation of therapeutic heating in some parts of tumor is considered a minimum goal for effective therapy. Ideally, one should treat normal tissue to tolerance, and temperature monitoring in normal tissue also should have a high priority.

In many early clinical applications, temperatures were measured at one or two points in tumor, usually in the nodule center. As more serious efforts were made to characterize clinical heatings, it became clear that such sparse sampling is not representative of how well tissues (tumor and normal) are heated. The results of these more aggressive assessments of thermal distributions reveal considerable temperature heterogeneity even in small target volumes. Figure 1.1 is an illustration of a planar temperature distribution reconstructed from invasive temperature measurements made during a microwave (MW) treatment of a superficial lesion. Temperatures range over several degrees within the target volume. The distribution in Fig. 1.1a might be predicted based on knowledge of

the energy deposition pattern associated with the MW applicator. It has a shallow central hot spot that is the result of the cool surface temperature and MW attenuation with tissue depth. In contrast, Fig. 1.1b is less easily related to expected energy deposition pattern. The lateral hot spot may be due to a region of poorly perfused or necrotic tissue. It may also be the result of an anomaly in the energy deposition pattern (e.g., an edge effect) of the MW applicator. Pronounced temperature heterogeneities make it clear that considerable advances in either noninvasive thermal imaging or mathematical modeling will be required in order to completely or even semicompletely quantify clinical treatments, particularly for deeply seated lesions. Nevertheless, correlations between invasive sampling and tumor responses have been reported by several investigators [5, 9, 36, 37], and the continued use of invasive thermometry is clearly indicated.

1.2.1 Practical Applications

It is worthwhile describing in some detail representative examples of thermometry applied to clinical treatments. Consider two situations:

1. superficial tumors, which are more accessible to invasive temperature probe placements, and
2. deep-seated tumors, which are generally more difficult to approach via a percutaneous probe placement and involve greater risk to the patient.

In the case of the superficial treatment, the target volume should be identified in a quantitative way

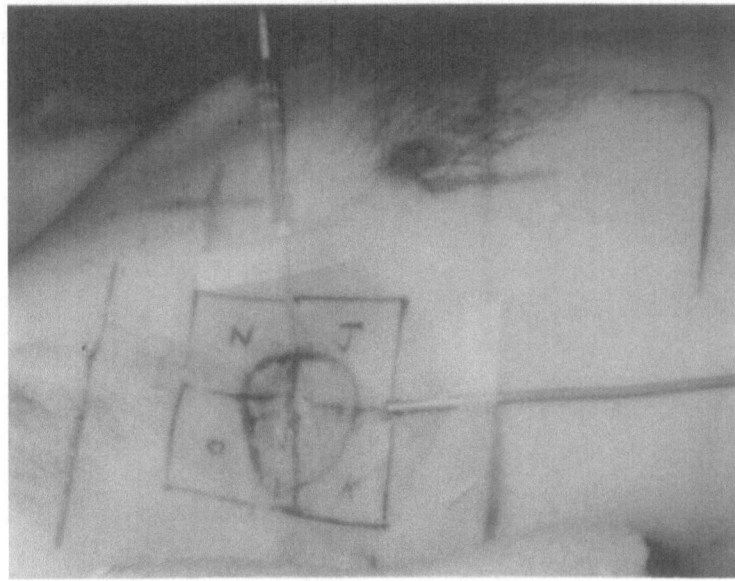

Fig. 1.2. Skin markings outline the target region and the projection of the heating applicator orientation prior to a treatment. A photograph documenting temperature probe positions is kept as part of the patient's treatment record

(measure length, width, and depth) using an imaging technique such as US, computed tomography (CT), or magnetic resonance (MR), or by palpation. It is also useful to outline the surface projection of the lesion with a marking pencil on the skin (Fig. 1.2).

The positions of the invasive probes are chosen based on several factors. It is desirable to monitor temperature extremes (i.e., minimum and maximum temperature). The minimum tumor temperature has been shown to have prognostic value with regard to therapeutic response [5], whereas the maximum temperature in normal tissue can have significance with regard to normal tissue toxicity. These extremes are determined by power deposition and heat loss via conduction and blood flow. In the final analysis, probes must be placed according to suitability of the invasive track. Consideration of vital structures (e.g., carotid artery in the case of a neck nodule) often determines how and where a probe can be placed, irrespective of the priority for temperature monitoring.

In superficial cutaneous or subcutaneous nodules, blood flow is often restricted in the nodule center. Thus, the central regions of such surface lesions are often sites of highest temperature. Power deposition from external unfocused superficial heating devices is generally highest in the center of the device aperture, and is attenuated with depth into tissue. In addition, blood flow and conduction heat loss are often high at the lateral and deep tumor edges. Thus, minimum temperatures are expected at deep lateral edges of the lesion and these regions should be monitored. If normal skin covers the lesion, this normal tissue may be the treatment-limiting factor. Surface sensors on the anticipated hot spots in normal tissue (e.g., skin in the central field, surgical scar) are essential for treatment without the complications of blisters or burns.

Documentation of the locations of measurements positions can be accomplished by defining a local coordinate system with respect to the skin surface. Coordinates x and y might represent positions on the skin surface, with the z coordinate measuring depth below the surface. Photographs of the treatment field are essential for reconstructing the treatment setup (Fig. 1.2).

Descriptions of the measurement positions (e.g., probe #1 is at the lateral tumor edge at 2.0 cm depth) should be documented in the treatment record. Even in cases of superficial lesions it is not easy to determine whether the invasive measurement sites are located in tumor or normal tissue. Radiographs, US, or CT scans may demonstrate that the actual location of measurement is not where it was expected or desired. An example of this is illustrated in Figure 1.3. The intent was to position three probes in the target volume defined by the CT image of this large axillary tumor. The CT verification demonstrates the misalignment in the actual probe placements. Two of the probes lie in the superficial tumor-normal tissue border. The third probe was intended to traverse the central diameter of the tumor. It also is displaced towards the superficial tumor edge. Therefore, pre- or post-treatment radiographic localization films are useful for defining accurately and documenting the location of invasive probes. This documentation process is time consuming and costly. Detailed documentation of probe location may be restricted to the first treatment setup in the treatment series if the probes or catheters are left in place. Another alternative is to reinsert probes or catheters close to the original locations for subsequent treatments by carefully recording the initial insertion points, angles, and depths.

For more deeply seated lesions the problems of invasive measurements are compounded. Because of

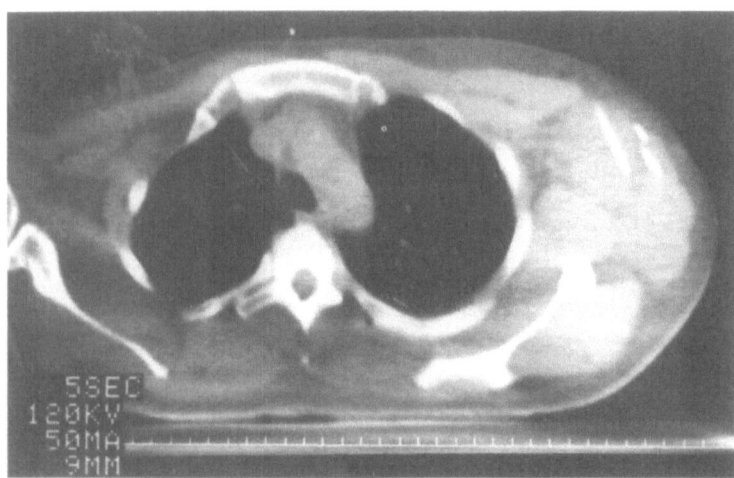

Fig. 1.3. A CT documenting the position of three invasive temperature probes. The intent was to place these probes deeply into the axillary tumor volume. In fact, two probes are superficial to the CT-defined tumor volume at the normal tissue-tumor interface and the third probe is much more superficial than the pre-CT estimate

technical limitations with respect to power delivery at depths greater than a few centimeters, it is very difficult to anticipate where temperature extremes might occur when treating deep-seated lesions. Regions of tumor necrosis, regions that appear hypodense on contrast-enhanced CT scans, usually have the highest documented temperature [38]. The CT scan in Fig. 1.4a illustrates a large pelvic recurrence with a hypodense central region. This hypodense central zone is the region of highest temperature (Fig. 1.4b). The volume at the tumor-normal tissue interface often shows steep thermal gradients and often is the region of lowest observed temperature (Fig. 1.4b) [38].

Without the ability to visualize or palpate the target volume, careful preplanning of the invasive track must be done before placing probes in deep-seated tissues. The geometry of planned probe or catheter placements can be determined using CT or MR images. It is often advantageous to insert the probe or catheter with CT, US or X-ray film guidance. In all cases it is important to verify the probe location. This can be done with repeat CT scans (Fig. 1.5a). When using CT verification the most convenient approach is to place the probe or catheter track in a single CT cut, a "radial placement" (Fig. 1.5a). Radial placements, however, are not always practical or indicated. In placements with an axial component the probe or catheter location can be verified using a multiplanar reconstruction (Fig. 1.5b). Alternatively, orthogonal or stereo X-ray films often suffice (see Fig. 1.5c). Although time consuming, the orthogonal film pair with original target-defining CT scans can be used to reconstruct the true track location. This is similar to the common practice of locating the positions of interstitial sources in radiation brachytherapy (Fig.

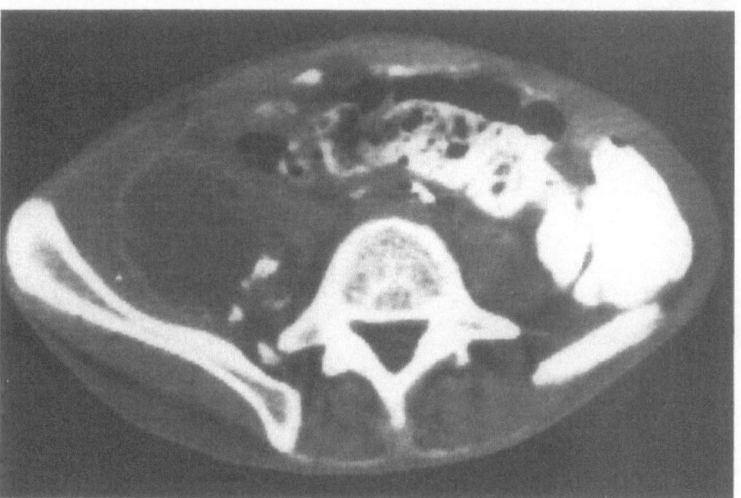

a

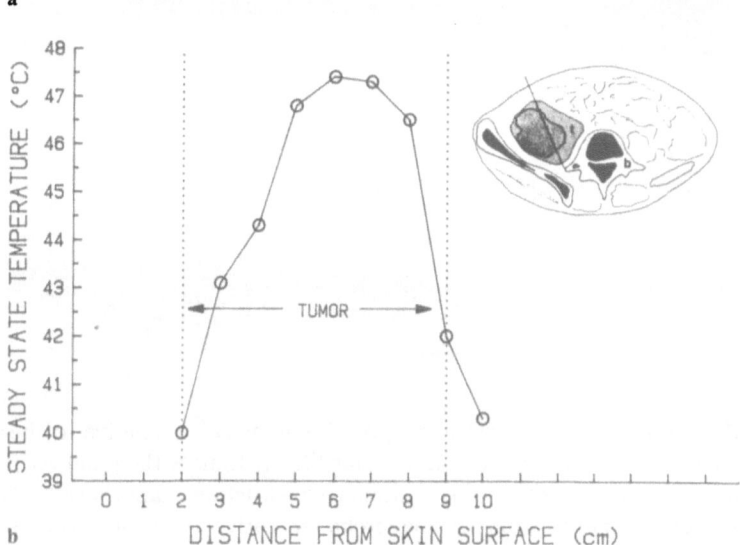

Fig. 1.4. a A contrast-enhanced CT scan of a large pelvic recurrence showing a hypodense central region in the tumor volume. b A temperature profile measured during a clinical treatment. The profile was obtained by mechanically incrementing a single-sensor temperature probe along an indwelling catheter traversing the tumor volume. (The catheter track is indicated by the *line* across the tumor in b.). (Bagshaw et al. [41])

b

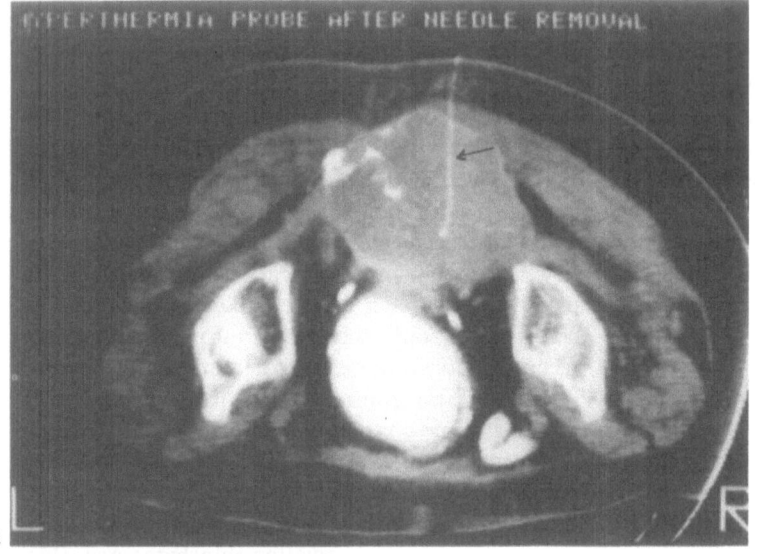

Fig. 1.5. a Position verification by CT of a temperature probe catheter (indicated by *arrow*) using a radial placement. **b** Multiplanar CT reconstructions (*middle* and *right figures*) for position verification of two temperature probe catheters (indicated by *arrows*). **c** Orthogonal films used to define the position of temperature probe catheters and interstitial heating electrodes for a combination hyperthermia-brachytherapy treatment of a base of tongue lesion. (Prionas et al. [49])

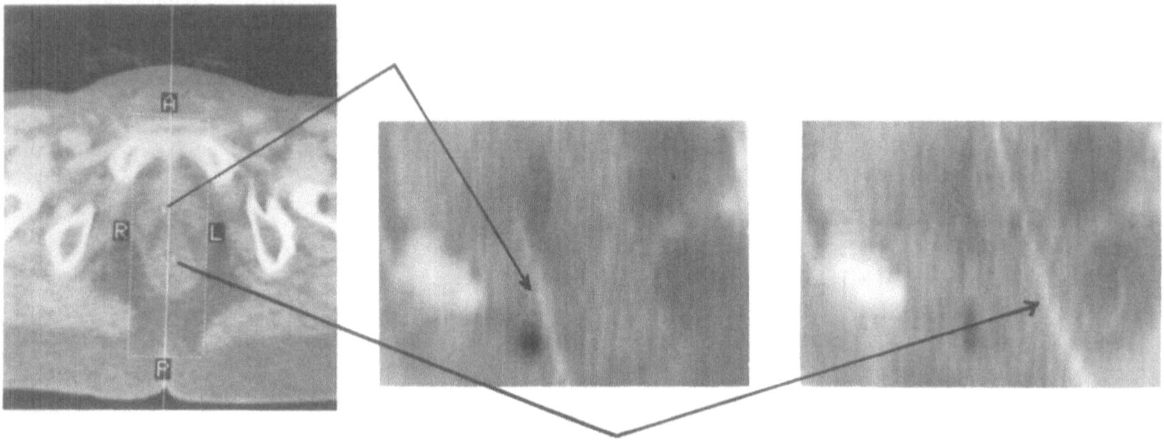

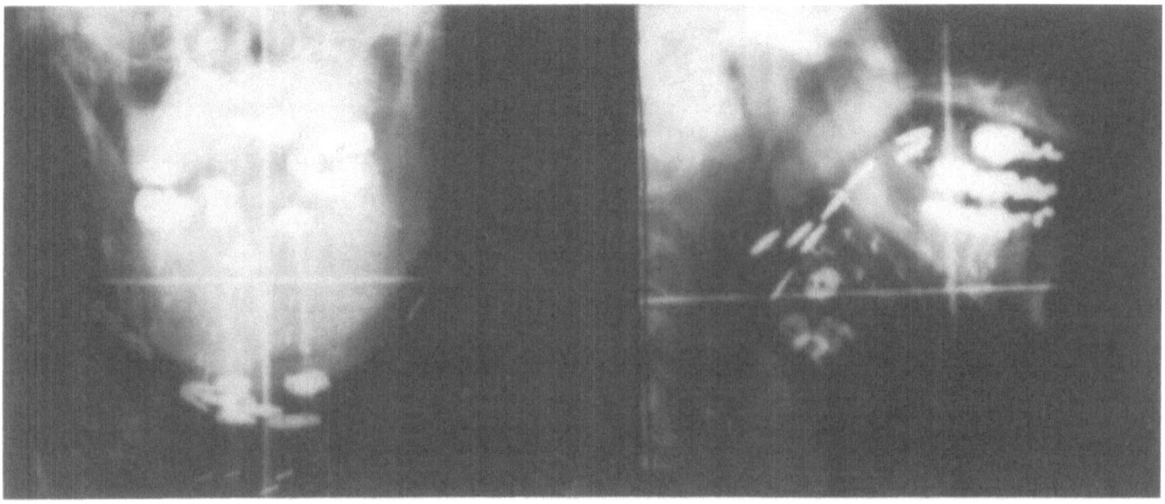

1.5c). Anatomical or externally placed landmarks, which can be identified on both the plane X-ray films and CT scan, are used for this position reconstruction procedure. The position of the probe or catheter relative to these landmarks is determined on the orthogonal verification films and then transferred to the appropriate CT slice. When working in the pelvis, the femur heads are a convenient set of landmarks. A line

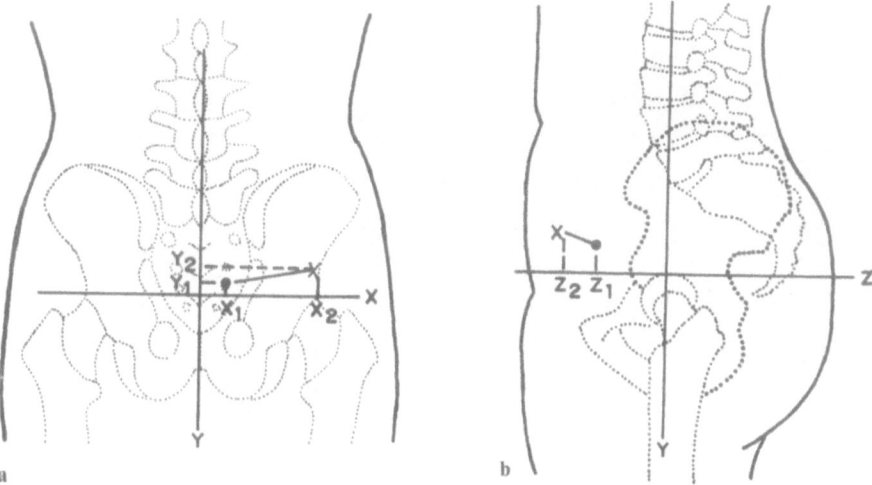

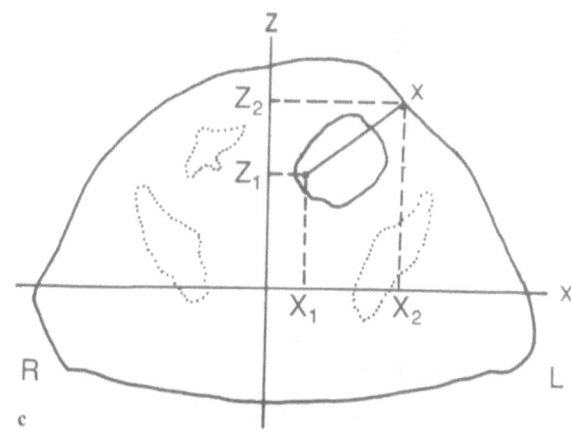

Fig. 1.6a–c. This figure illustrates the procedure of localizing the position of a percutaneously placed catheter using orthogonal X-ray films and a diagnostic CT scan. An orthogonal reference frame is chosen based on anatomical landmarks. In the case of the pelvic region the femur heads serve this function (**a, b**). A radiopaque catheter, or a catheter with a metallic insert, can be identified on the anterior and lateral X-ray films. The end points of the catheter track (● indicates the catheter tip at depth and × indicates the catheter at the skin surface) are located and coordinates (x, y, z) are identified with respect to the reference frame. These positions (appropriately corrected for film magnification) can be transferred to an equivalent reference system with respect to the CT scan(s) (**c**). In this illustration a single slice is adequate to locate the y positions of both the catheter tip (●) and skin entrance point (×). The lateral (×) and anterior-posterior (z) position of the tip (●) and entrance (×) points can be used to establish the catheter track in the CT slice

drawn horizontally at the top and center of the femur heads, and vertically through the pubic symphysis and spinal column, defines the origin of an orthogonal coordinate system. This system, identified for both the plane film and CT scans, can be used to localize the position of temperature probes in both imaging techniques (Fig. 1.6).

A variety of techniques are used for inserting interstitial temperature probes. A sterile technique is recommended (Fig. 1.7), the specifics of which can depend on the heating modality. For electromagnetic heating a temperature probe and catheter combination is often used and the probes need not be sterile [39]. In the case of US, plastic catheters produce excessive local heating and their use is not recommended. Bare metal inserts such as thermocouple needles are the most common probe type for this heating modality. In this case the actual probe has to be sterilized.

In the probe-catheter combination used for EM heating, a thin-wall biocompatible plastic catheter with a bore diameter that fits snugly about the temperature probe is first introduced into the tissue at the desired location. Special closed-end hard-tip catheters have been designed for this purpose (Best Industries, 7643-B Fullerton Rd., Springfield, VA; Deseret Med Inc., Park Davis & Co., Sandy, UT 84070). After the skin has been pierced or lanced, these catheters can penetrate tissue when forced with a metal stylet that provides strength and rigidity (Fig. 1.8). Alternatively, insertion can be accomplished using a multistep procedure (Fig. 1.9): First, a metal trocar or breakaway needle with an internal bore of sufficient size to accommodate the plastic catheter is introduced into the tissue at the appropriate site (Fig. 1.9-1). Once the trocar is in place, the catheter is inserted (Fig. 1.9-2), and the trocar is then removed by sliding it back over the catheter (Fig. 1.9-3). [If the trocar is a breakaway needle (Fig. 1.7), it can be separated and removed without cutting the catheter hub.] This leaves the catheter in place of the metal trocar (Fig. 1.9-4). Using an open-end angiocatheter is another option; however,

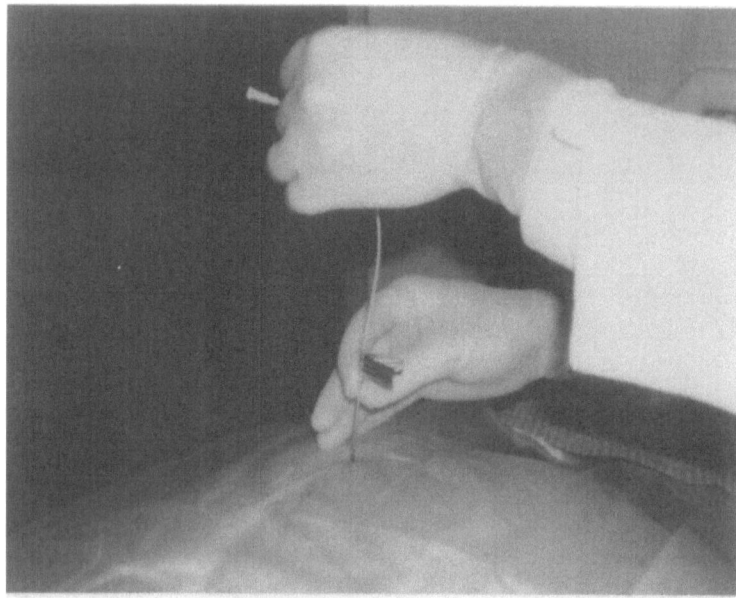

Fig. 1.7. Photograph of a sterile technique used for a catheter insertion via the breakaway needle technique

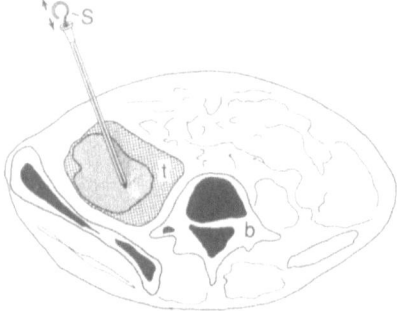

Fig. 1.8. The percutaneous placement of the hard-tip catheter. At the time of insertion the catheter is reinforced with a metal stylet (s). The stylet is removed and replaced by the temperature probe. The tumor region is indicated by "t", "b" represents bone

catheters should be checked for slippage and infection. In many cases it is necessary to remove the original catheter and reinsert a fresh one because of signs of wound infection or catheter slippage.

An indwelling catheter provides a track in which a temperature probe can be inserted and, as originally proposed by Gibbs [34], the probe can be moved incrementally within the catheter, allowing temperature sampling along the length. This process of "temperature mapping" has been automated and has greatly increased the data acquired during treatment sessions, providing a larger data base for analysis in terms of heating quality and prognostic thermal parameters [3, 9, 40]. A demonstration of such a temperature map is shown in Fig. 1.4b [41].

In the case of US, instead of the plastic catheter, a metal needle may be used through which a probe can be moved to obtain similar temperature profiles. The metal needle reduces the selective absorption heating that occurs with the plastic catheter materials. However, viscous heating of the needle remains a possible artifact and the use of a metal shaft introduces the possibility of thermal smearing due to the metal's relatively large (compared to plastic or tissue) thermal conductivity. This may result in a false impression of temperature uniformity. The combination of a 20-gauge spinal needle and a small (0.5 mm o.d.) fiberoptic probe has minimal thermal smearing artifact. Another disadvantage of metallic needles is the fact that it is usually not possible to leave these rigid needles in place throughout the course of treatment.

extra precaution is needed in this case since the open-end catheter is a possible source of infection. The closed-end catheter technique is preferable since, with care, the catheter can be left in place for the entire treatment course (which may last several weeks), and alleviates the need for sterilizing the temperature probe. If the option to leave the catheter in place for the treatment course duration is chosen, it is useful to insert a plastic filler between treatments to prevent the catheter from having any acute bends that will prohibit the subsequent reinsertion of a temperature probe. Also, the catheters need to be secured by tape or suture to prevent slippage due to tissue movement. Topical antibiotics can be applied at the percutaneous entrance to minimize the risk of infection. At each treatment session, or more often if necessary, the

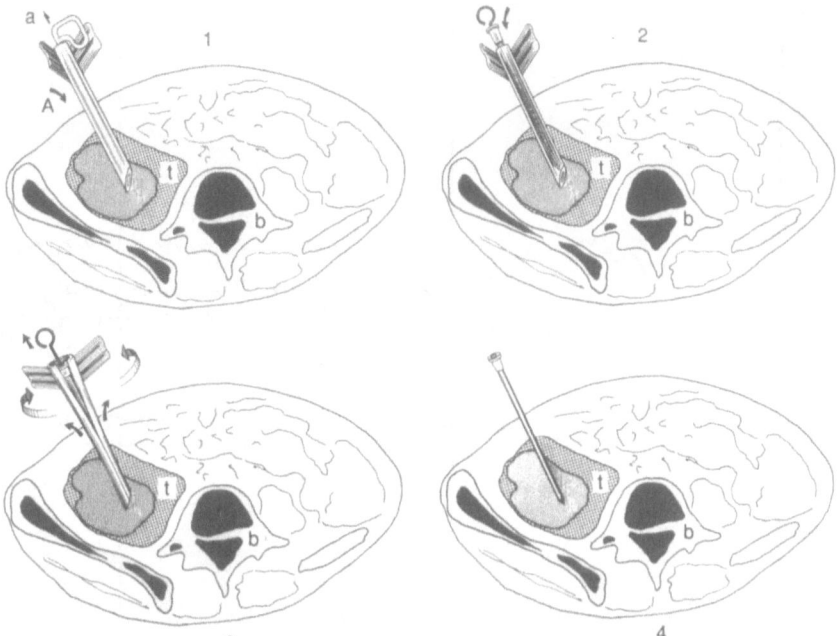

Fig. 1.9. A schematic illustration of the insertion steps (*1–4*) associated with flexible soft-tip catheters using a breakaway needle (Deseret Med Inc., Park Davis & Co., Sandy, UT 84070) ("t" and "b" are as in Fig. 1.8.)

1.2.2 Temperature Control

In the current state of the art, active control of the temperature distribution is very limited. First generation equipment designs are simplistic and do not address adequately the complexity of clinical heating. Often this equipment was designed and tested on phantom models or anesthesized animals. These models do not simulate adequately the actual clinical therapy. The static phantom models obviously lack the means to account for blood flow heat loss. Although the anesthetized large animal models (e.g., pigs) are a better approximation, they too are inadequate because perfusion in the anesthetized subject is usually reduced markedly. In fact, animals under prolonged anesthesia have problems keeping a stable core temperature, and blood flow in the extremities and superficial tissue is reduced in order to maintain temperature in deeper vital organs. These in vivo heating simulations give an erroneous impression of how well a given device or system can respond to blood flow dynamics and achieve temperature uniformity in a given volume; thus, the complexity associated with real-time temperature control in human subjects is underestimated.

Most attempts at real-time temperature control use a single feedback temperature sensor. The location and specification of the sensor is chosen by the therapy technologist. Once chosen, maximum and minimum temperature limits can also be defined by the user.

The basic control algorithm is one that modulates the overall heating device power in order to maintain this control sensor in the desired temperature range. One problem with this control scheme is that the control point is somewhat arbitrary and generally cannot be assumed to be representative of other temperatures in the treatment field. Temperature data are often displayed during therapy in real time as a temperature vs time plot. Such a plot is shown in Fig. 1.10. Each curve traced on the graph represents the temperature at a single measurement point as a function of time during the treatment course. Generally, this time course can be separated into three regions:

1. the heat-up, indicated by a rapid rise in temperature commencing with the application of power to the heating apparatus;
2. a steady state, during which temperatures remain relatively constant or show slow variations in time as a result of changes in power or blood flow (vasodilatation);
3. the cool-down, indicated by the rapid fall of temperature to near the initial baseline value.

The spatial variations in temperature are readily apparent from the spread in the temperature traces (see Fig. 1.10). Thus, controlling a single point temperature has questionable relevance to the quality of the treatment.

More sophisticated efforts to control temperatures have been proposed for simple fixed-power distribu-

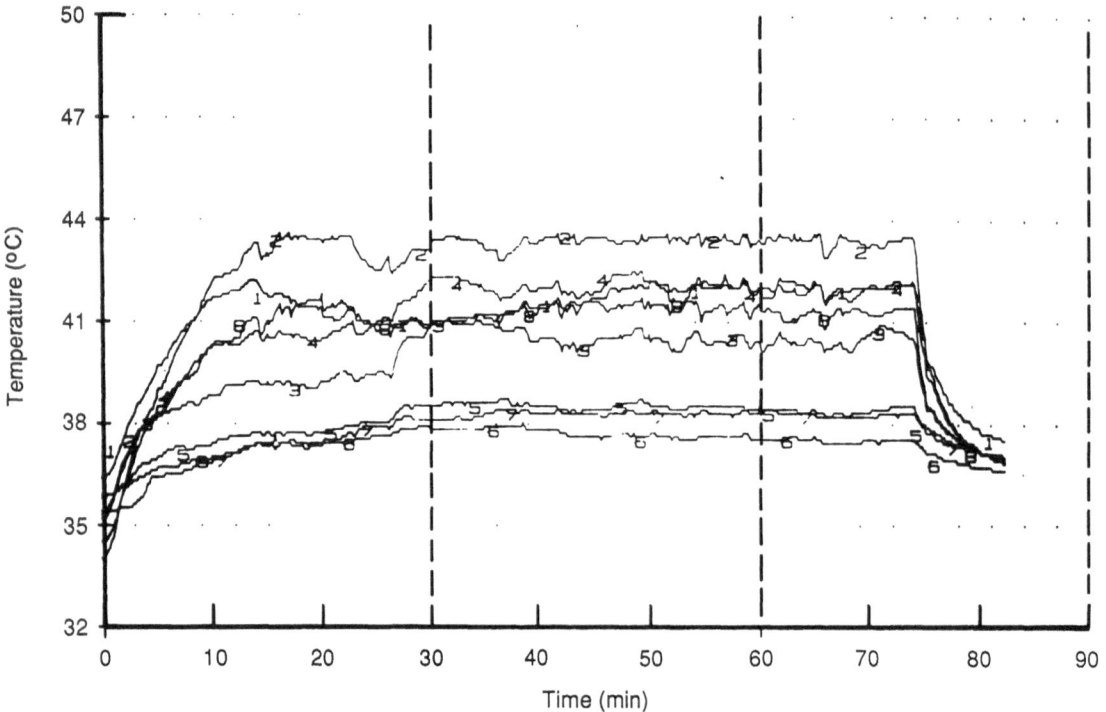

Fig. 1.10. An example of a typical temperature vs time plot associated with a clinical hyperthermia treatment

tion devices. A two-point control algorithm has been formally worked out by Knudsen and Heinzl [42]. In this approach the two points are controlled by adjusting device power and surface temperature. Since the general temperature distribution for most fixed-field superficial heating devices has a predetermined form (i.e., cooler surface temperature, maximum temperature at some depth, then decreasing temperatures with increasing depth in tissue), such a two-point control approach has merit. If one determines that the two control points defined by probe placements lie on either side of the temperature maximum, it can be inferred that all points on a line between the two control points are above the chosen control temperatures. In this way one extends temperature control over a region instead of to a single point. However, in order to verify that the distribution between the control points defines a local maximum, a multisensor array probe or linear map may have to be employed.

Equipment designs that have the capability to manipulate the local power deposition in a given target volume of tissue offer an enhanced capability for temperature control. These include devices that employ scanning (e.g., scanned focused US [43, 44]) and/or multielement power deposition sources (e.g., array radiators of either the MW or US type and interstitial heating elements) [45–47]. These techniques require that the positions of temperature sensors be well defined with respect to the geometry of the heating device in order to have spatially consistent feedback from the temperature sensors to the various power sources. Figure 1.11 illustrates the effect of a closed-loop control algorithm associated with a multitransducer US system (Labthermics Sonotherm 1000 system. Labthermics Technologies, Inc., 701 Devonshire Dr., Champaign, IL 61820). After the initial rapid temperature rise, computer control is turned on (~7-min mark) for two of the probes and corresponding US transducers. Power to these respective transducers is modulated in order to keep these temperature readings at the predetermined constant value. Two additional probes, which are not under computer control, continue to measure a rise in temperature with time. At approximately 10 min these latter two probes are also switched to the automatic control algorithm. During the control period (~10–40 min) all probes are maintained at nearly stable temperatures via power modulation of the appropriate US transducers. At 40 min into the heating session, the temperature control system is disengaged and temperatures are allowed to drift at constant power levels or with intermittent manual power adjustments.

These more sophisticated approaches to control the temperature distribution require a larger number of

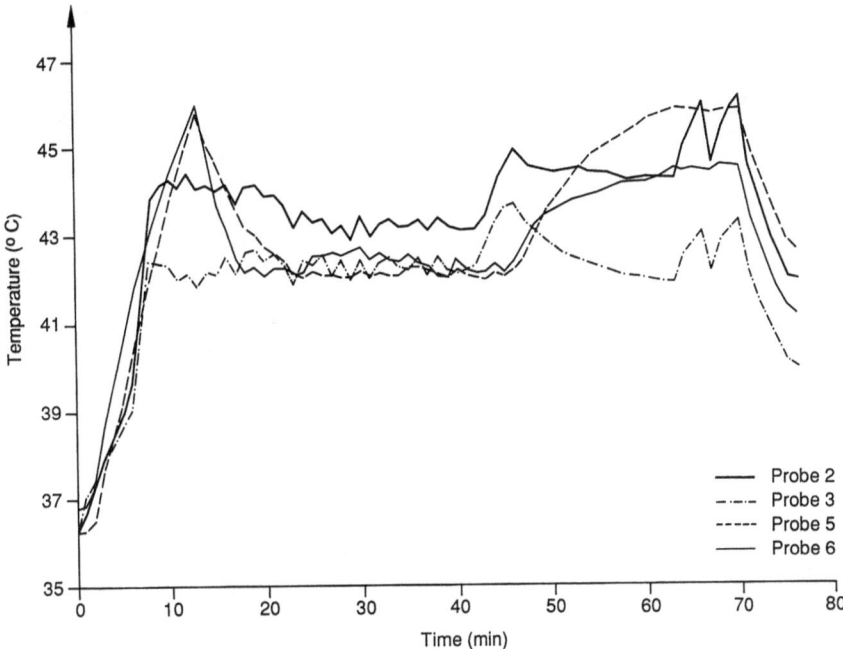

Fig. 1.11. An example of the temperature vs time plot obtained during a hyperthermia treatment using a multisector US applicator (Labthermics Technologies Inc., 701 Devonshire Dr., Champaign, IL 61820). In the time interval between 10 and 40 min, four temperature probes are under automatic control and temperatures remain at relatively stable values except for the fluctuations associated with power modulations to the respective transducer sectors. At approximately 40 min the autocontrol algorithm is turned off and temperatures are allowed to respond to manual power manipulations

temperature sensors. In practice the number of invasive probes that can be inserted into the target volume can be restricted easily. Therefore, the number of adjustable parameters available for temperature control with a given heating device may be limited by the current state of invasive temperature monitoring. An exception may be the interstitial heating techniques. With these approaches the necessary feedback control temperature sensors can be incorporated conveniently into the interstitial heating electrodes or antennas (Fig. 1.12) [46, 48–50]. Four to five sensors spaced along each invasive track in a typical interstitial radiation/hyperthermia implant array constitute a reasonable grid of temperature data for control purposes. Such data allow the systematic control of temperatures in various regions of the grid to desired levels by regulating power or other relevant parameters in the corresponding heating element(s). The limitations to implementation of this type of controlled heating are the current lack of cost-effective multichannel thermometry systems having 20–100 artifact-free thermometer channels and the need to develop computer control algorithms for this large number of temperature sensors.

Perhaps the most important limitation in attempts to control the temperature distribution during hyper-

thermia stems from the fact that information other than temperature often determines the treatment course. Most notable is patient discomfort. Complaints of pain override monitored tissue temperatures when controlling a treatment. In an analysis of 996 treatments with numerous heating devices and modalities used on various anatomical sites, Kapp et al. reported that 526 treatments were associated with pain symptoms, with 136 treatments being limited to temperatures less than 42.5 °C largely due to pain symptoms [3]. Methods for controlling pain are certainly available; however, there is the risk of treatment-related morbidity if this important source of patient feedback is ignored. Cautious use of drugs for pain control and improved temperature monitoring may be advisable. However, it is important that such steps be carried out systematically with respect to both adding pain modifiers and temperature control in order to quantitatively demonstrate any gains that are made in therapeutic heating.

Finally, there is some potential for controlling superficial hyperthermia treatments using noninvasive thermometry. The simplest approach is to use a large number of skin surface temperature sensors, since skin is the normal tissue that should be protected from heat-induced toxicity. It is also the site of most

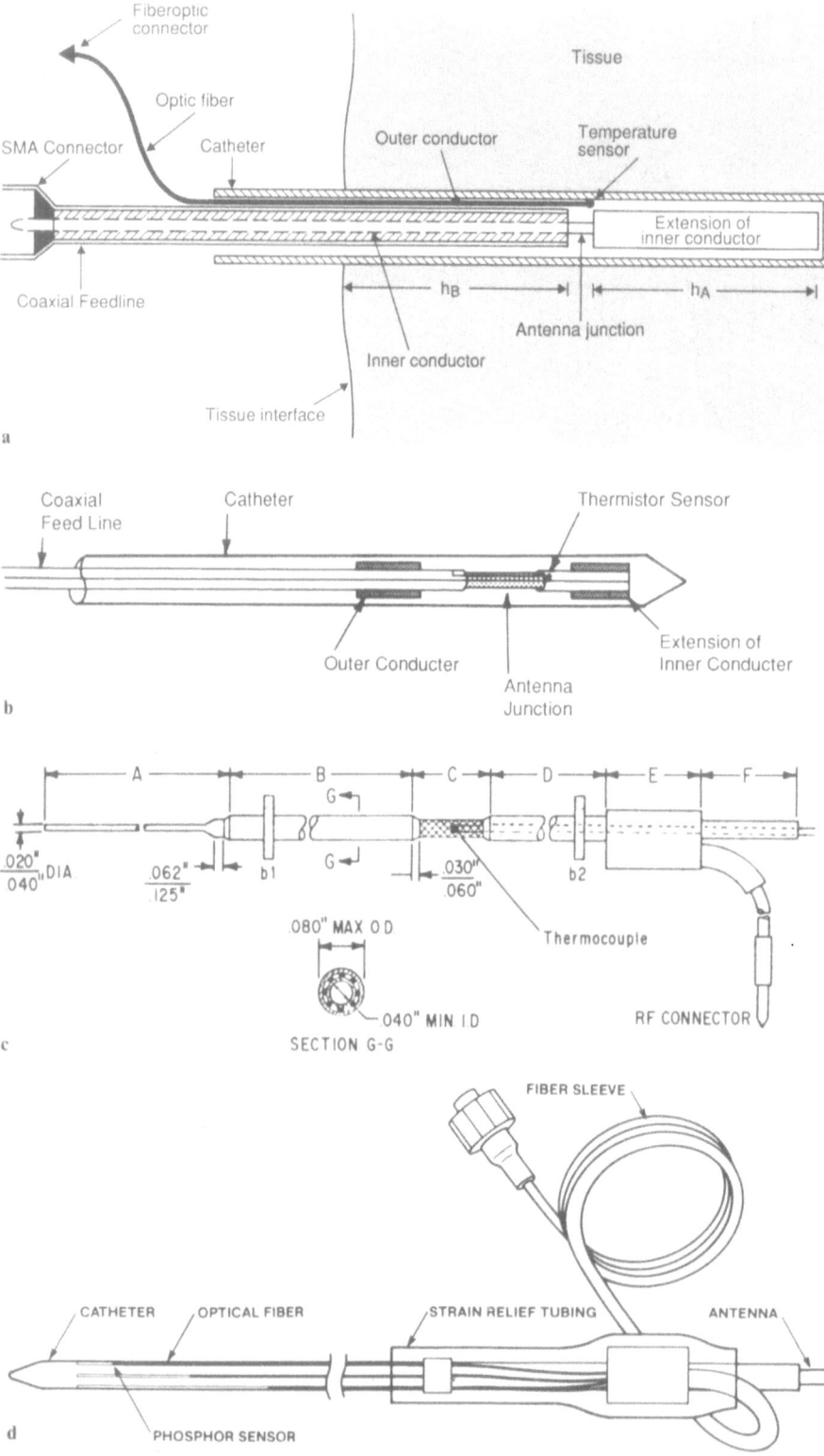

Fig. 1.12a–d. Four examples of combining temperature monitoring and interstitial heating elements. **a** A MW antenna with an adjacent fiberoptic probe placed in the same nylon catheter (Strohbehn and Mechling [46]). **b** Thermistor sensor integrated into a MW antenna between the inner and outer con-ducting elements of the coaxial antenna (Turner [48]). **c** A ther-mocouple sensor inserted inside a flexible braided RF electrode (Kapp et al. [3]). **d** A MW antenna catheter with several fiberop-tic sensors integrated into the catheter wall. (Sun and Samulski [50])

pain nerve receptors. Therefore, treating the skin to tolerance in terms of toxicity and pain is the most aggressive way to treat the underlying superficial lesion. Skin temperature can also be measured and controlled using infrared thermography. A prototype device that uses a noncontact scanned MW radiator and infrared sensor has been implemented clinically [51]. Alternatively, surface as well as subsurface temperature information can be obtained noninvasively via MW radiometry. An integrated MW heating antenna and radiometer has been developed and used for controlled therapy [52]. Optimization of superficial therapy using such techniques deserves further investigation. It may be possible to demonstrate clinically that the therapeutic efficacy achieved through these noninvasive control schemes is equivalent to that achieved with the more aggressive invasive temperature monitoring and control. However, if the interest is to quantify rigorously temperature distributions for the purpose of establishing the performance of heating equipment, or to document the advantages and disadvantages of heating techniques, invasive measurements are essential.

1.3 Available Technologies

1.3.1 Thermocouple Thermometry

Thermometry using thermocouples has several desirable advantages: the thermocouple junction is intrinsically calibrated depending only on the choice and quality of materials used to form the sensing junctions. The response of the thermocouple to temperature is nearly linear over the range of interest in hyperthermia (Fig. 1.13). The size is essentially unrestricted, allowing for multisensor arrays with dimensions and flexibility suitable for tissue implantation. The sensors are inexpensive, and support equipment including a digital voltmeter and multiplexing data acquisition system is readily available at reasonable cost.

Thermocouples derive their function from the temperature dependence of the Fermi level in the free-electron gas of a metal's conduction-energy band. This parameter is well above absolute zero at $300\,°K$ ($27\,°C$), and depends weakly on temperature. In practice the temperature dependence of the Fermi level is exploited by observing small changes in the contact potential (difference in electron work functions) between dissimilar metals at two separate junctions maintained at different temperatures. The tempera-

ture sensitivity of the difference in contact potential between two dissimilar joined metals is on the order of $10-60\,\mu V$ per degree. Achieving accuracy and resolution of a tenth degree requires microvolt measurements, and although this may be done readily with today's technology, care is required to ensure that errors are not made. The intrinsically low signal level and the substantial difference in electrical and thermal conductivity with respect to tissue are the major disadvantages of thermocouple temperature probes.

A diagram of a basic thermocouple system is shown in Fig. 1.14. Two different metal components, copper and constantan, are in contact at two separate junction points (J_1 and J_2). One junction is maintained at a known stable reference temperature T_2 and the measured voltage is related to the temperature difference ($T_1 - T_2$) between the reference temperature and that of the measurement junction T_1. For standardized thermocouple materials (e.g., copper-constantan) the reference temperature is chosen to be the water ice point and a universal voltage versus temperature response curve can be measured [53] (Fig. 1.13). That is, the same absolute voltage will be measured for a given metal pair with a given temperature difference between J_1 and J_2. Thus, the junction is intrinsically calibrated. It is worth emphasizing that use of intrinsically calibrated probes has the advantages of allowing probe interchangeability and reducing the possibility of the large measurement errors that might result if probes were used without calibration checks. In Fig.1.15 the distributions of temperature differences between thermocouple probes and a calibrat-

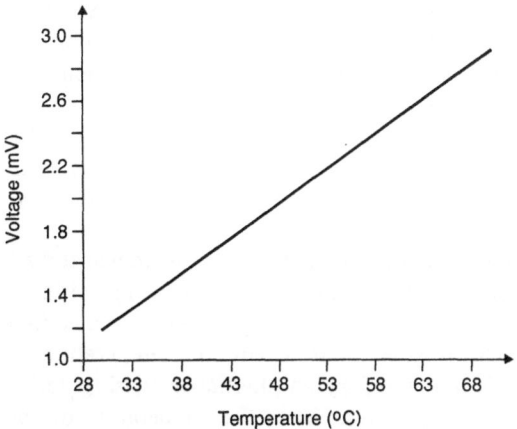

Fig. 1.13. The universal voltage vs temperature response for the copper-constantan thermocouple over the temperature range of interest in hyperthermia

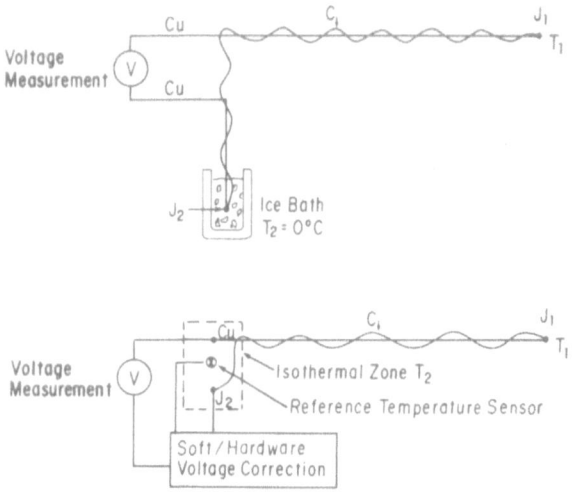

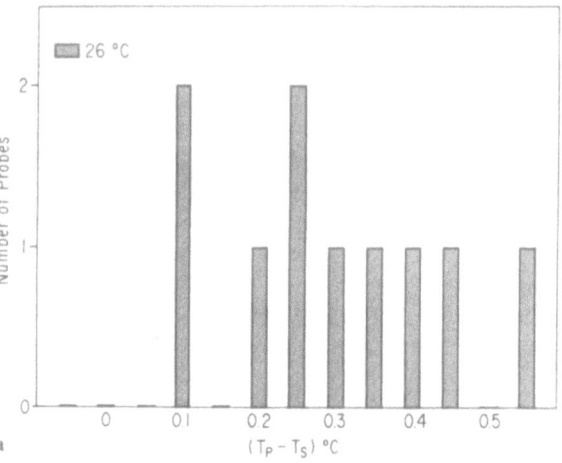

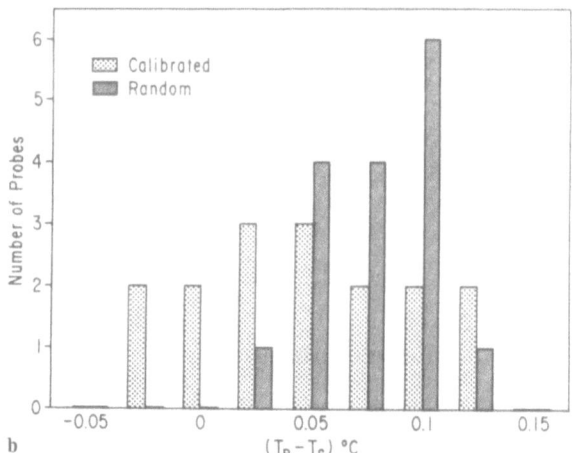

Fig. 1.14. Schematic representation of the basic copper (*Cu*)-constantan (*C*) thermocouple sensor with a water ice bath temperature for the reference junction (J_2, *upper diagram*) and a software or hardware compensated reference junction (J_2, *lower diagram*)

Fig. 1.15a, b. An example of the distribution of thermocouple probe deviations from a standard thermistor ($T_p - T_s$) for a clinically used thermocouple thermometry system. **a** Distribution of probe differences ($T_p - T_s$) at 26°C prior to calibrating the system. **b** Distribution of probe differences after system calibration (calibrated) and after a subsequent random interchange of the probes with respect to the thermocouple input channels (random)

ed standard thermometer are plotted. In Fig. 1.15a the distribution is that associated with ten randomly chosen probes connected to an uncalibrated commercial thermocouple thermometry system. The largest deviation is on the order of 0.5°C. This range of calibration variance is typical for properly functioning thermometry systems and probes. In a multi-institution survey of thermocouple thermometry, Shrivastava et al. found the standard deviation of differences between thermocouple probe readings and a standard thermistor thermometer reading to be 0.32°C based on 641 probe evaluations [54]. Figure 1.15b illustrates the accuracy associated with thermocouple probe interchangeability. The data in this histogram were obtained with a 16-channel thermocouple thermometry system, which was calibrated with the probes connected in one configuration and then subsequently redistributed randomly among the various channels. Although there is a systematic shift in the probe difference distribution before and after randomization, the range of probe errors remains well within ±0.2°C.

An ice bath is often an inconvenient reference and an electronic substitute may be employed (Fig. 1.14). In this configuration the second junction is situated in a thermally isolated environment, the temperature of which is monitored by an alternative sensing device such as a thermistor. When the temperature of the second junction is known, one can correct back to the standard reference temperature (ice bath) by adding or subtracting an offset voltage. This correction can be made via a software algorithm (software compensated) in computerized data acquisition or by means of an electronic circuit (hardware compensated). Although convenient, using an electronic ice-point substitute introduces the possibility for calibration error since one relies necessarily on the accuracy of this independent reference sensor and accompanying compensation. Usually, a single-point calibration of such electronically compensated thermocouple systems is sufficient to ensure the recommended 0.1°–0.2°C accuracy for clinical hyperthermia applications. An example of the temperature errors observed after a single-point system calibration is illustrated in Fig. 1.16. The system calibration point was at the lowest temperature, 26.0 °C. As seen in the figure, the spread in the error distribution increased at temperatures displaced from the calibration point (35.5 °C, 43.9 °C), but the overall range in errors remained within satisfactory limits for clinical use.

Although any two metals can be used to form thermocouple junctions, a number of material combinations have been characterized that are useful for various applications. The copper-constantan thermocouple junction is perhaps used most commonly in hyperthermia applications, since the sensitivity and linearity are adequate in the temperature range of interest (Fig. 1.13). Other junction types such as manganin-constantan have been suggested to reduce thermal conductivity of the metals and minimize the electrical conductivity imbalance in the junction [25]. This latter consideration may be important in reducing selective heating in the thermocouple junction if such a probe is used for measurement during EM heating.

Thermocouple thermometry systems are readily available from several commercial sources since this technology has widespread application. Systems for use specifically in clinical hyperthermia are also available (Labthermics Technologies Inc., 701 Devonshire Dr., Champaign, IL 61820), which offer some software (quick single point calibration) and hardware (RF shielding) features that are preferable in hyperthermia applications.

1.3.2 Electrical Resistance Thermometry

The change in electrical resistivity of a material with temperature is a well-known phenomenon. For metals, electrical resistivity increases with increasing temperature. However, metal resistivity is generally low in absolute terms (e.g., 1.6×10^{-8} Ω-m for silver, 10.7×10^{-8} Ω-m for platinum), and the associated temperature coefficients are also small (+0.00392 Ω/ Ω − °C for chemically pure platinum). These very low changes in resistivity with temperature changes limit the use of the metallic resistance temperature detector (RTD) in hyperthermia. For example, the metallic RTD cannot easily be made physically small enough for use as an interstitial probe. In addition, in an applied situation small changes in the resistance of wires or contact points in the resistance measurement circuit can introduce substantial errors in the inferred temperature. Therefore, RTD devices are not employed commonly in hyperthermia except as calibration standards [26]. With appropriate measurement techniques and instrumentation, the platinum

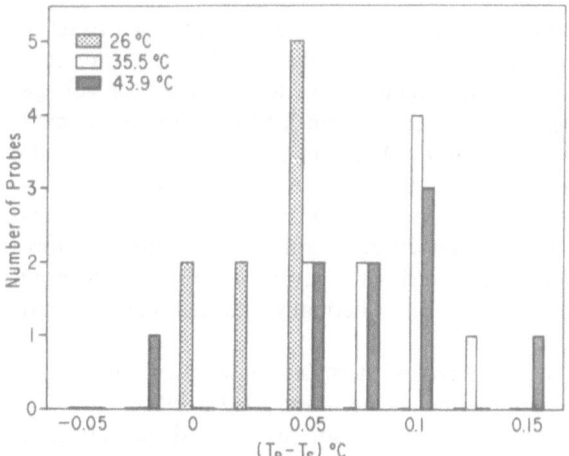

Fig. 1.16. An example of the distribution of thermocouple probe differences $(T_p - T_s)$ from a thermistor standard at several temperature points (26°, 35.5°, 43.9 °C) after the single-point system calibration at 26 °C

RTD offers a means of highly accurate temperature measurement over an extremely wide range of temperature and is used as the interpolation standard between the oxygen melting point (−182.96 °C) and the antimony melting point (630.74 °C).

Thermistor-based thermometry overcomes the physical size and resistance measurement problems associated with the metallic RTD. It also offers a significant improvement in sensitivity when compared to the thermocouple. With a thermistor, the temperature-dependent parameter is the resistivity of a semiconductor material. This resistivity is inversely proportional to the number of charge carriers (electrons in the conduction band and holes in the valence band) that are mobile under the influence of an applied electric potential. In a nondegenerate material the Fermi level lies in the energy band gap (E_g), a few units of kT_a (k = Boltzmann's constant, T_a = absolute temperature) above the valence band and below the conduction band. Thus, the number of electrons that can make the energy gap transition (valence to conduction band) is strongly dependent on temperature since this number is proportional to e^{-E_g/kT_a}. Typical changes in resistance for thermistors are on the order of 4% per degree Celsius. Therefore, the careful measurement of semiconductor resistance can be a sensitive indicator of temperature.

Thermistor sensors are usually made from various metal oxides such as nickel, iron, and manganese, and can easily be made physically small enough for interstitial probe applications. With available stable current sources, the more accurate technique for making the resistance determination is the use of a four-lead arrangement (Fig. 1.17). Two leads are used to drive a constant small current through the thermistor sensor and the second pair of leads is used for voltage measurement. The current level should be kept small (approximately 10 µA) in order to prevent Joule heating within the sensor. A measurement at two different current levels can be made to determine the extent of sensor self-heating if this is a potential problem. Also, reversing the current direction can be used to eliminate any thermally induced contact emf voltage [26].

The advantages of thermistor-based thermometry are high resolution and accuracy, good stability, small size allowing multiple sensor arrays, and cost-effective technology. Thermistors can also be adapted for use in EM fields. This last advantage is made possible by using the technique of measuring the thermistor resistance via high-resistance leads [11]. A high-resistance material (carbon-impregnated Teflon) with

a resistance similar to tissue will prevent the induction of large currents, which are the source of EM artifact and EM field perturbations associated with metallic leads. Thermal conduction smearing is also minimized with this configuration because the high-resistance leads have low thermal conductivity. However, probe length and diameter reduction are limited with these high-resistance lead probes. Consequently, multisensor arrays are not available. The high-resistance lead thermistor configuration has been integrated into commercial hyperthermia systems (BSD Corp., 520 Chipeta Way, Salt Lake City, UT 84108). Standard thermistor configurations are also available in clinical hyperthermia systems (Humanoid Systems, 17022 Montarero St., Carson, CA 90746).

One disadvantage associated with thermistor probes is the lack of intrinsic calibration. This generally means that each sensor has to be calibrated individually to achieve desired accuracy. A potential for large measurement errors exists if thermistor probes are interchanged or inappropriately selected in multichannel thermometry systems. In addition, the resistivity is not a linear function of temperature (Fig. 1.18), and three or more calibration points are needed to characterize the response curve [55, 56]. The extra effort required in thermistor calibration procedures is, however, offset by their good long-term calibration stability.

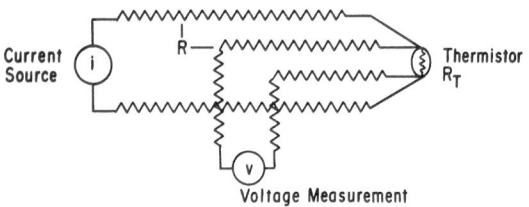

Fig. 1.17. Schematic of a thermistor resistance measurement circuit. A small constant current is driven through the thermistor (R_T) and the voltage drop proportional to the thermistor resistance is measured at contact points adjacent to the thermistor sensor. If the circuit's lead resistances(R) are large (e.g., 100 kΩ/cm), induced currents in these leads will be small and selective heating of the probe in EM fields will be minimized. (Bowman [11])

1.3.3 Gallium Arsenide Optical Thermometry

As indicated, the electrical properties (i.e., resistivity) of semiconductors have a high degree of temperature dependence. Similarly, optical absorption in semiconductor materials can also have pronounced temperature dependence. For optical photon energies less than the valence-conduction band gap energy (E_g), a semiconductor material will be essentially transparent. However, for photon energies greater than or equal to E_g, the material will be highly absorbing. This transition between transparent and opaque is very sharply defined by the gap energy and occurs over a relatively narrow range of photon wavelengths. The edge absorption transition can be very sensitive to the temperature of the semiconductor, since this light absorption process can be assisted by lattice vibrations (phonons) and depends on the creation of excitons (bound states of the electron-hole pair). In some materials (e.g., GaAs) the absorption can be empirically expressed in the form:

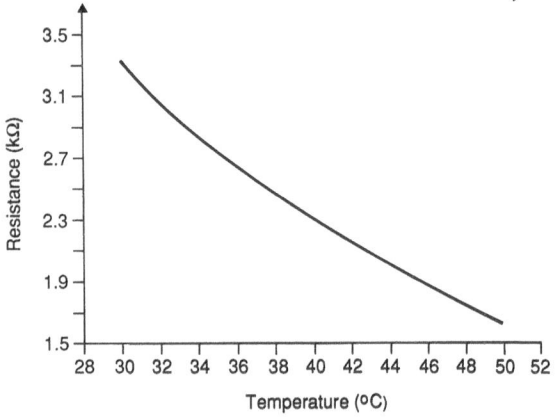

Fig. 1.18. Typical resistance vs temperature response for a thermistor sensor in the temperature range of interest in hyperthermia

$$A = A_0 \exp\left[\alpha\left(h\nu - E_g\right)/k\,T_a\right] \tag{1.1}$$

where α is a temperature-insensitive material-dependent constant and h is Planck's constant. Thus, for a fixed light frequency with energy (hν) close to E_g the absorption will be strongly dependent on the thermal energy (kT_a) (Fig. 1.19). Christensen designed an optical temperature probe based on the band edge absorption shift of GaAs [13]. A schematic of this optical thermometry system is illustrated in Fig. 1.20. The basic design is one in which a narrow band light source, generated by a light emitting diode (LED), is transmitted down one optic fiber channel through a small reflecting prism of GaAs and returned through a second fiberoptic channel to a photodiode detector. The light transmission through the GaAs prism is related to the prism's temperature as shown in Fig. 1.19. The solid-state light source and detector are very cost-effective components; however, they require operating temperature stabilization since their photoelectric efficiencies (particularly for the LED) are also temperature dependent. This thermometry technology is employed commercially and multichannel (12 channels per thermometry system) systems are available with the Clinitherm hyperthermia systems (Clinitherm Corp., 12046 Forest Gate Dr., Dallas, TX 75243).

The GaAs optical thermometry concept has several advantages, the most important being functional immunity in the presence of strong EM fields. In addition, the optical sensing element is small and can be configured in multisensor arrays with size amenable to interstitial implantation [22]. The probe materials are plastic and therefore minimize errors associated with thermal smearing.

There are several drawbacks to the clinical use of this thermometry system. First, the sensors are not intrinsically calibrated and the sensor response is not linear, requiring a number (10–20) of calibration points in order to characterize the response curve. The probes are also susceptible to strain artifacts. For example, bending the optical fibers will result in reduced optical transmission in the fibers that cannot be distinguished from temperature-induced GaAs absorption

changes. Two other cautions should be noted with regard to this type of thermometry. One is that water strongly absorbs light at the GaAs band edge frequency, and moisture absorption in plastic optical fibers can affect optical transmission, resulting in a calibration error. Watertight encapsulation is necessary and the integrity of this waterproofing needs to be insured for clinical use. Secondly, this thermometry is limited to temperatures below 50°C, which may not be high enough for some interstitial hyperthermia applications. Finally, the plastic materials in the probe are subject to US absorption artifact.

1.3.4 Photoluminescent Thermometry

Thermometry technologies based on photoluminescent materials have also been developed specifically for EM-induced hyperthermia (Luxtron Corp., 1060

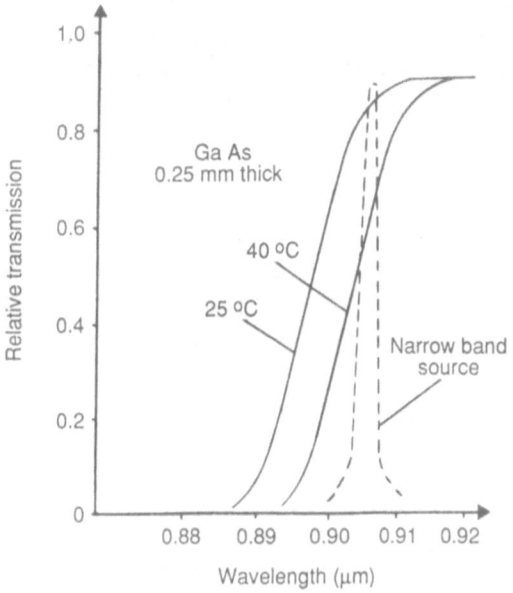

Fig. 1.19. The relative optical transmission of GaAs as a function of wavelength for a 0.25-mm sample at two temperatures (25°, 40°C). (Vaguine et al. [22])

Fig. 1.20. Schematic of the GaAs optical thermometer. Light emitted from the LED is propagated along an optical (transmit) fiber through the GaAs crystal prism and back through another optical (receive) fiber where the intensity is measured with the photodiode detector (PD). Light absorption in the GaAs prism is temperature dependent; thus the measured intensity at the PD can be related to the GaAs crystal temperature

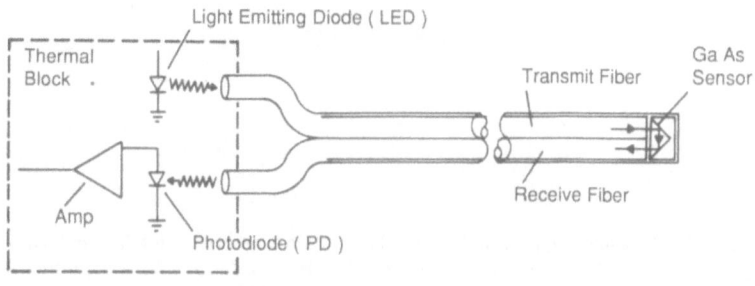

Terra Bella Ave., Mountain View, CA 94040). These technologies utilize the temperature-sensitive properties of luminescent materials and, like the GaAs system, feature probes having a fiberoptic channel with a thermally sensitive element attached at the end. In general, photoluminescent materials absorb light in one frequency range (the excitation spectrum) and subsequently emit light in a different (usually lower) frequency range (the emission spectrum). The phenomenon, therefore, involves at least two different wavelength bands in the optical domain which can be separated via a spectral filtering technique. One advantage gained from this frequency separation of the excitation and emission light signals is that a single optical fiber can be used for a probe configuration. However, the photoluminescent efficiency of most materials is relatively low and excitation signals are generally orders of magnitude larger than emission signals. Thus, the spectral isolation has to be very good in order to compensate for this large difference in signal intensity.

First generation photoluminescent thermometry systems (Luxtron models 1000 and 2000 models) are based on the temperature dependence of photolumines-

cent efficiency (i.e., the relative change in the emission intensity for a fixed excitation intensity as a function of temperature) [15]. In practice a material with several narrow emission bands is used. Since each band quenches (luminescent efficiency decreases) at a different rate as a function of temperature, the ratio of the light intensity in two emission bands is in principle an intrinsic temperature-dependent parameter for the chosen material. These emission intensity ratios can therefore be used to determine the temperature in the material. A schematic of the measurement instrument is shown in Fig. 1.21. A tungsten-halogen lamp is used as the excitation light source. Light from this lamp is filtered to pass ultraviolet (UV) wavelengths, and focused into the optical aperture of the quartz fiberoptic probe. The UV light transmits down the quartz fiber and excites the multiband fluorescent sensor. The fluorescent emissions propagate back through the same fiber and are focused and split into two optical paths. These two optical paths are filtered separately, allowing the transmission of the desired emission bands. The intensities of these different emission wavelengths are detected with two separate photodiode detectors. The outputs of the

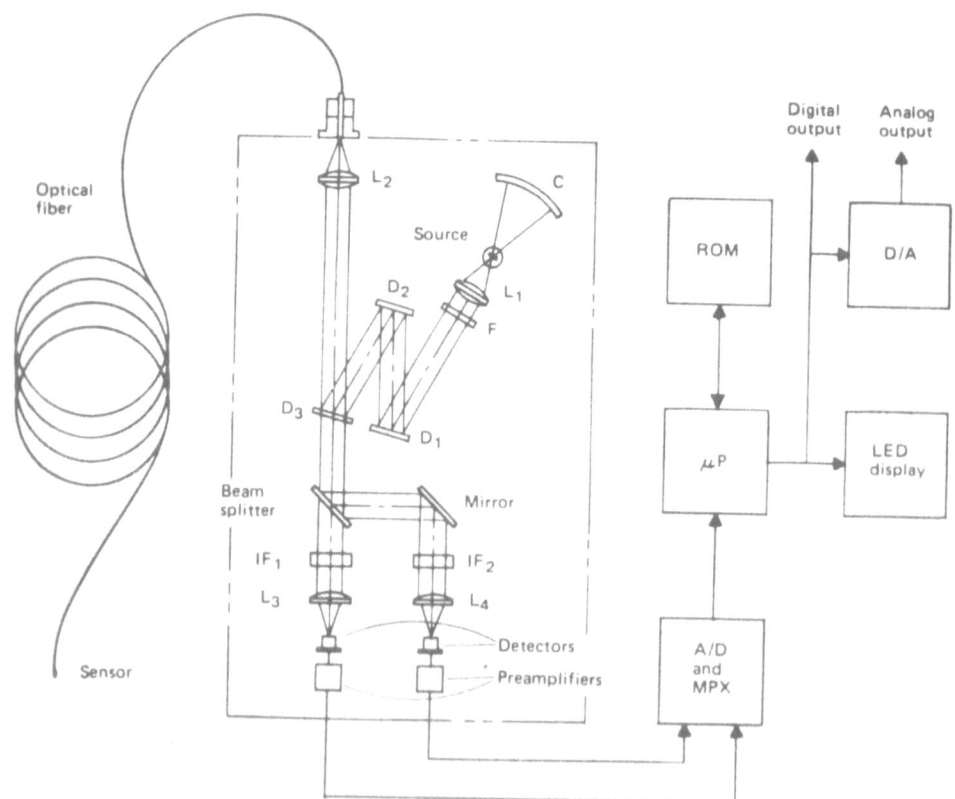

Fig. 1.21. Schematic representation of the photoluminescent thermometry system (Luxtron 1000 B and 2000 B) based on determining temperatures using the ratio of two different luminescent emission wavelengths. This technique requires a fairly sophisticated optics module requiring a precise geometric alignment. (Wickersheim and Alves [15])

diode detectors are fed through a multiplexer to an analog to digital converter and to a microprocessor. The ratio of the signals is calculated and compared with a stored look-up table using the microprocessor, and the temperature determination is output or displayed. The look-up table values are adjusted to give an accurate temperature reading based on a two-point calibration procedure. Since this system is operated using a continuous rather than a modulated excitation source and uses two independent diode detectors, there is the potential for error to result from DC offset drift in the detectors. In order to correct for this source of error, the system, under microprocessor control, periodically shuts off the tungsten lamp and measures the detector background signal levels. These levels are then used to correct the fluorescent signal levels in order to calculate an accurate signal ratio. The Luxtron 1000 and 2000 model series use this emission intensity ratio technique with the sensitive material being europium-doped lanthanum oxysulfide (Fig. 1.22). This material is excited with UV light, and thereby the probes require high-quality glass optical fibers for good transmission of the UV wavelength. Also, the emission bands are relatively narrow, being on the order of 100 Å, and thus require high-quality spectral filters for proper wavelength separation. Since the filter transmission can change as a function of temperature, these and other optical components have to be thermally regulated to obtain adequate system performance.

Photoluminescent thermometry based on the spectrum ratio technique has the advantages of probes that are free of artifact in an EM field and that, being constructed of glass and plastic materials, do not perturb the EM field. These probes also are low in thermal conductivity. The ratio detection technique renders this measurement of temperature insensitive to variation in the excitation lamp intensity. However, the intrinsic calibration potential of these systems could not be realized because of practical technical limitations. For example, bending the optical fiber differentially affects the fiberoptic transmission of the different emission frequency bands. Thus, the respective intensity ratios can be misinterpreted as a temperature change if the fiber is bent sufficiently. Temperature dependency in the spectrometer components (such as the narrow bandpass optical filters) also compromises system performance, resulting in drift in the system calibration. Although the technology can meet the performance goals outlined for clinical hyperthermia, the quality assurance maintenance required for these systems makes them somewhat inconvenient for routine clinical use.

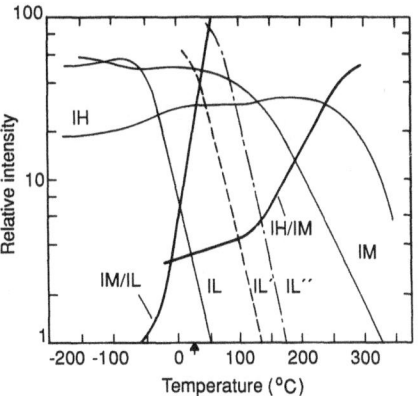

Fig. 1.22. Temperature quenching curves (*IH*, high; *IM*, medium; *IL*, low) for europium emission lines in different oxysulfide hosts. *IL* represents intensity of one line in lanthanum oxysulfide; *IL'* is for same line in gadolinium oxysulfide; *IL''* refers to yttrium oxysulfide. Ratios of two different pairs of europium emission in lanthanum oxysulfide are also shown (*IM/IL* and *IH/IM*). (Wickersheim and Alves [15])

A second generation photoluminescent thermometry system (Luxtron model 3000) has recently been developed for use in hyperthermia [23]. This system is based on the temperature dependence of the luminescent decay time, the rate constant associated with fluorescent or phosphorescent afterglow. In principle, the decay rate constant is an intrinsic property of a given photoluminescent material and thus its dependence on temperature should be an intrinsic material property. Probes associated with the system, as presently developed, appear to be universally calibrated to within a degree or two (Fig. 1.23). Although this is not sufficient to allow probes to be used interchangeably, it can prevent large measurement errors when probes are not correctly calibrated or are inadvertently interchanged. In addition to this near universal calibration response, the dependence of the decay rate constant for the chosen material (magnesium fluorogermanate) is linear in the range from 0° to 80°C, and a single point calibration in the hyperthermia range (30°–60°C) is sufficient to achieve the calibration performance goal. Using an intrinsic optical property of a single emission line also greatly reduces probe sensitivity to bending stress. Plastic fiberoptic temperature probes of 0.5 mm in diameter have demonstrated temperature changes on the order of 0.1°C when subjected to a bend radius of 0.5 cm [23].

In practice, the sensitive luminescent element is attached to a single optical fiber having an outside diameter of approximately 200–500 µm (Fig. 1.24). This element is excited via a filtered xenon flash lamp

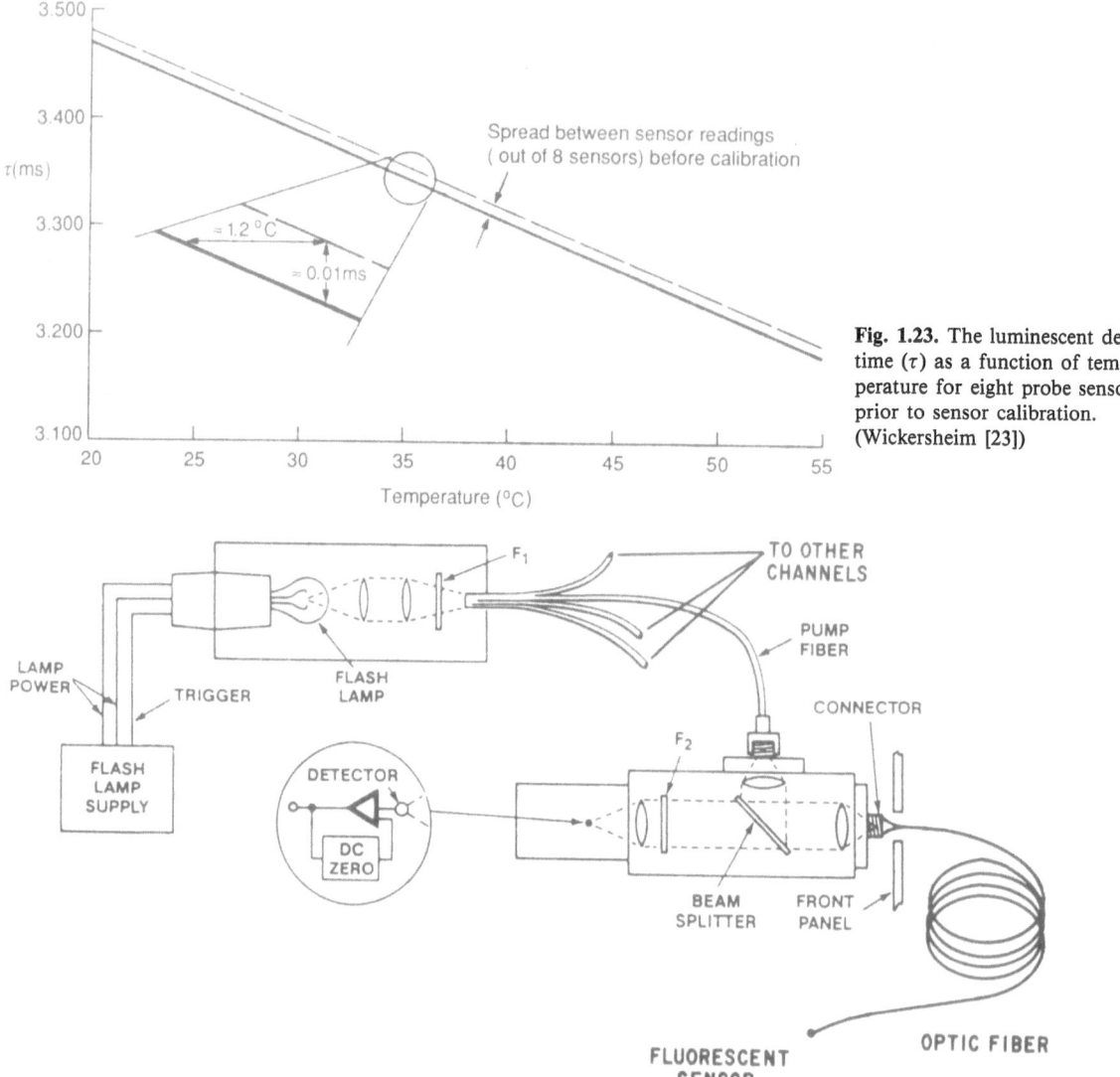

Fig. 1.23. The luminescent decay time (τ) as a function of temperature for eight probe sensors prior to sensor calibration. (Wickersheim [23])

Fig. 1.24. Schematic of the photoluminescent thermometry system (Luxtron 3000) based on determining the luminescent decay time. A single filtered xenon flash lamp is used as a source to several of the system's probe channels. In each separate channel the flash is reflected by a beam splitter and focused into the fiberoptc link which is terminated with the fluorescent sensor. The resultant fluorescent light emission progagates back down the same fiber and is detected using a photodiode. The time rate of change in the fluorescent emission is used to determine the sensor temperature. (Wickersheim [23])

with light pulses that are of microsecond duration. The luminescent decay time for the sensitive material is in the millisecond range. Thus, excitation and emission signals are well separated in the time domain. The process through which the temperature-dependent decay time (τ) is determined is illustrated in Fig. 1.25. This illustration is an idealization and the actual decaying fluorescent signal has additional random noise superimposed. Therefore, a number of pulses (approximately 20) have to be averaged in order to obtain the temperature accuracy and resolution required in hyperthermia. In the currently available systems,

the xenon flash lamp can operate at a nominal rate of 10 flashes per second. Thus, a measurement accurate to $\pm 0.2\,°C$ can be made in approximately 2 s. Better accuracy and resolution can be obtained by averaging a larger number of flash pulses. This, in turn, increases the measurement time. In most situations data acquired at a 2- to 5-s rate will be adequate for clinical applications.

The positive features of this luminescent decay time thermometry are EM immunity, low probe thermal conductivity, insensitivity to bend artifact, probe size less than 1 mm in diameter (allowing array configura-

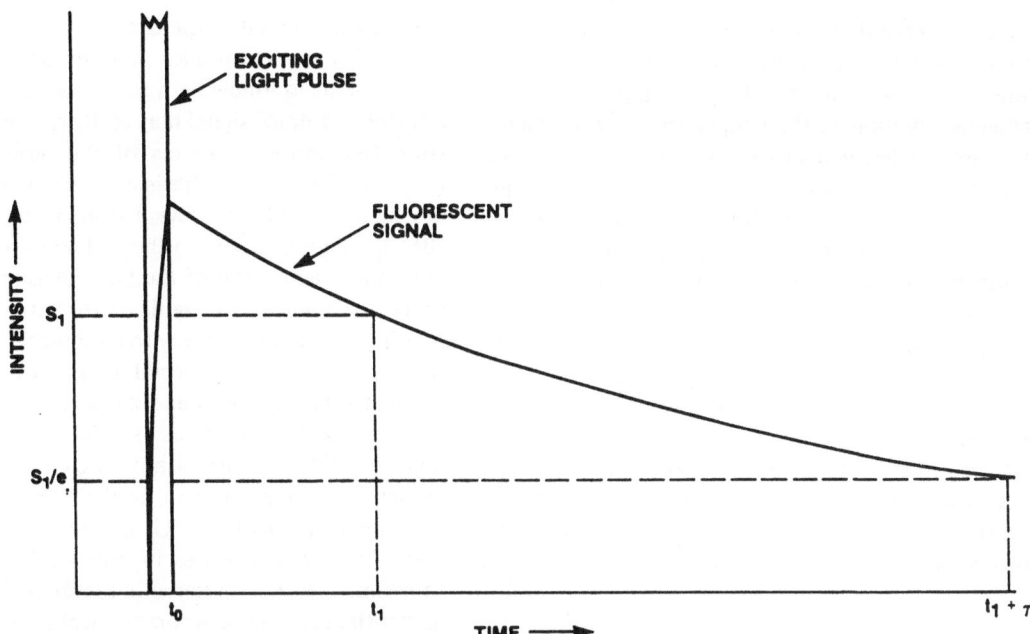

Fig. 1.25. The process by which the fluorescent decay time is measured. A short high-intensity light pulse is used to excite the fluorescent sensor at time t_0. A short time later (t_1), after the electronics recover from the high-intensity pulse, the level (S_1) of the fluorescent signal is determined and a clock circuit is initiated. When the signal detected at t_1 decays to $1/e$ of its value S_1, the clock is stopped and the decay time (τ) is determined. (Wickersheim [23])

tions), and the single point calibration useful over the range of 0°–80°C. There are also some practical technical advantages over the fluoroptic ratio approach. The fluorescent material is excited at wavelengths in the visible spectrum so plastic optical fibers are adequate for fabricating probes. The emission spectrum is broad and narrow band filters are not required. Also, only a single photodiode detector is required; thus the system is technically simpler in terms of function and fabrication when compared with the spectral ratio approach.

One limitation in the current technology is that a single flash lamp serves as the excitation source for several probe channels. This lamp has to a accommodate a range of probe excitation levels since each channel may vary in optical efficiency due to differences in probe fabrication, optical coupling in the fiberoptic connectors, geometric alignment, and degradation due to probe use. Although the decay time constant is largely independent of the excitation intensity, the signal from various probes may have different signal to noise levels because of the above-mentioned factors. Thus, each probe requires a different optimum excitation intensity level. During the system calibration procedure, the flash intensity is automatically set to accommodate the signal requirements of the several channels. However, in some cases, it may not be possible to find a flash intensity level that will keep the probe signals from the several channels within the limits of the electronic amplifier gain associated with the photodiode detectors. When this happens, one or more of the probe channels that fall outside the gain limits will fail the calibration routine. Alternatively, even though all probes are successfully calibrated, since they may not be optimally excited in terms of flash intensity, the probes may perform with different levels of measurement noise and resolution. Software incorporated in the current systems automatically excludes probe channels that exceed a preset measurement resolution of 0.2°C. The above problems associated with the single excitation source can generally be resolved by replacing the temperature probes. The current probes are available as four-sensor arrays or as four single sensors attached to one four-channel connector, and the replacement of probes can be an expensive way to correct one deficient sensor. Future innovations with regard to the excitation source or probe design should rectify these current limitations.

1.4 Measurement Errors and Artifacts

Temperature measurements made during hyperthermia are assumed to approximate (within a few tenths of a degree) the tissue temperature in close proximity to the measuring sensor when the sensor, probe, and catheter are not present. This assumption is fundamental, and allows the temperature measurements to be used to decide treatment strategies either during therapy (varying applied power and determining treatment duration) or between treatment sessions (changing treatment devices), and objectively quantify treatments in terms of thermal parameters (maximum, minimum, and average temperatures). Several sources of measurement error and artifact can invalidate this basic assumption. These include the thermometry performance problems, such as calibration drift, as well as errors or artifacts that arise due to the fact that the probe/catheter/sensors are not tissue equivalent in relevant physical parameters (e.g., thermal conductivity, electrical conductivity, and acoustic absorption).

1.4.1 Calibration and Drift

It is essential in the practice of clinical hyperthermia that facilities have available equipment and techniques for calibrating thermometer probes from various systems with a high level of assured accuracy. The basic requirements are:

1. several (at least three) traceable standard thermometers (mercury-in-glass-thermometers, stable thermistors),
2. a stable temperature reference source (high-flow circulating water bath, ice points, melting points [57, 58], or triple points [59],
3. techniques for comparing standard thermometers and clinically used probes, and
4. means for accurately transferring calibration data from the standard intercomparison to clinical hardware or software.

The simplest and perhaps least expensive configuration is the circulating water bath and mercury-in-glass thermometers(s). A typical thermally regulated circulating bath can usually maintain temperature in the bath to within 0.1 °C using a thermistor-referenced heating coil. Mercury-in-glass thermometers are commonly inscribed with 0.1 °C markings and are easily readable by eye to within 0.05 °C. (Techniques for the precise use of mercury-in-glass thermometers have been outlined by Wise and Soulen [60]. The disadvantages of this simple standard calibration configuration are several. The response time of the mercury-in-glass thermometer is relatively slow; therefore, it tends to average out rapid temperature fluctuations

that may be present in the circulating bath. However, the smaller clinical probes may respond sufficiently quickly so as to follow bath fluctuations, and an electronically sampled signal may easily capture a temperature fluctuation in excess of the desired goal of 0.1°–0.2 °C accuracy. This is more likely to happen at the higher end of the relevant temperature scale (e.g., 50 °C), where heat loss to the ambient air may result in frequent activation of the bath heating element in order to maintain a stable bath temperature. Insulating the bath is one way to minimize this problem. Increasing the overall thermal inertia of the system by using metal calibration blocks is also a suggested way to reduce calibration errors associated with bath fluctuations [27]. Another disadvantage of the mercury-in-glass thermometers can be their limited temperature range (typically 51 °C) at the high end of the scale. Since temperatures in excess of this level are sometimes reached in necrotic tumor regions or near interstitial electrodes, accuracy checks up to 60 °C are desirable. Finally, the reading by eye precludes automation of the calibration intercomparison by computer control, a feature that can make repeated testing over extended periods more convenient.

Procedures for assuring calibration accuracy have been detailed by Cetas [26], and data regarding the performance of commercial thermometry systems are increasingly available [54]. It is recommended that a record of thermometry performance be established that will allow one to quantify the level of measurement confidence for all clinically used systems. Repeated calibration checks and extended time drift tests are ways of quantifying performance. An example of a drift test is illustrated in Fig. 1.26. Two probes are traced in the figure. One is the thermistor standard, the second is the clinical probe under test. The variations in the standard thermistor are largely the result of ±0.05 °C cycling of the water bath. A compilation of calibration data for a number of probe calibration checks for a clinical system are shown in Fig. 1.27. This accumulation of data is used to quantify the magnitude and frequency of calibration discrepancies. Tests such as these should be done prior to and following the initial introduction of a system to clinical use. Once consistency has been established and confidence in the system's performance gained, the frequency of checks can be reduced if indicated. Checking a new system pre- and posttreatment for calibration accuracy will give information related to probe durability when subjected to routine clinical use. It is a good practice to examine carefully all clinical temperature measurements and look for unusual readings such as high or low temperatures relative to the expected baseline temperature prior to

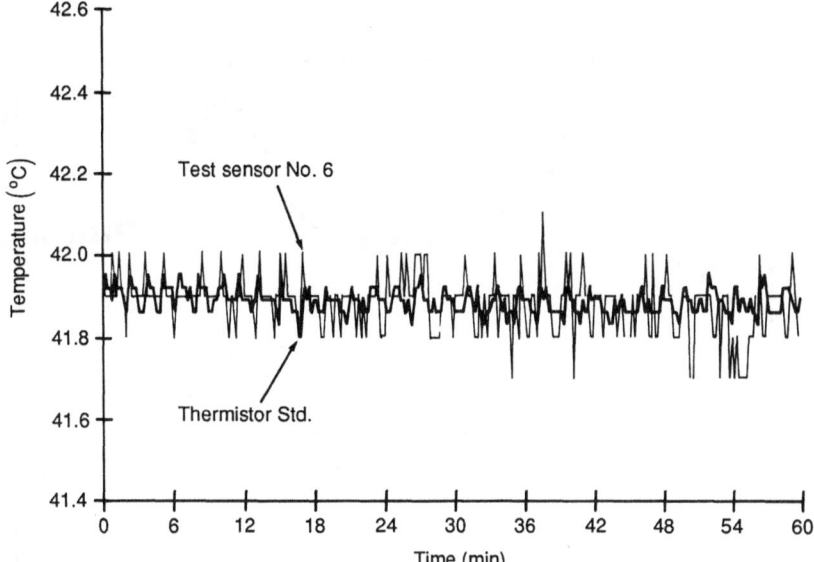

Fig. 1.26. The stability of a test probe is examined over a 1-h period in a stable water bath. The bath temperature is monitored by a thermistor standard (*thick trace*) and noise and drift of the test probe (*thin trace*) can be quantified by the standard deviation of readings and differences in average temperature measured at the beginning and end of the test

the start of treatment. Abrupt temperature changes during treatments associated with mechanical stress or probe movements may indicate performance problems. Finally, it is important to document all temperature measurement errors in the patient treatment record.

1.4.2 Thermal Smearing

An example of measurement error resulting from probe thermal conduction is shown in Fig. 1.28. In this example, a temperature profile has been measured in normal canine brain tissue close to an energized interstitial microwave antenna [61]. The profiles are measured using two different temperature probes. The discrepancy between the respective profiles is the result of the difference in the thermal conductivity of the probe shaft. One is a quartz glass optical probe, the other a stainless steel needle with an embedded thermistor sensor. Because of the large temperature field curvature near the interstitial device, temperature averaging or smearing associated with the metal needle results in a substantially lower maximum temperature as well as spatial distortion. It is important to realize that the thermal conductive properties of the probe and associated catheter can influence the measurement. The magnitude of the effect depends on probe/catheter physical properties, probe/catheter size, and the curvature of the temperature field that exists in the sampled tissue [62, 63].

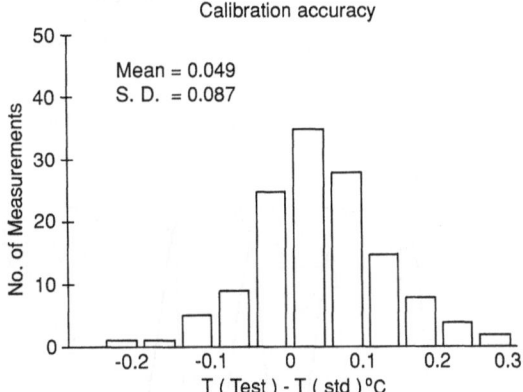

Fig. 1.27. A distribution of calibration comparisons for a thermometry system containing several probes is illustrated. Probes were calibrated and compared to a thermistor standard in a stable water bath. The differences between the test probes' and the standard thermistor's time-averaged readings for a period of 1 h are plotted in this histogram

One approach to quantifying the magnitude of thermal conduction effects on temperature measurements is to examine the deviation of a known temperature profile that results when this profile is measured using a test probe or probe/catheter configuration [61–64]. An example is shown in Fig. 1.29. In this case a nearly step-like (up and down) temperature profile is measured using two different temperature probes alternately in the same catheter. The probes are slowly incremented through the catheter tracing out the measured profile. One probe is a four-sensor array, plastic fiber, fluoroptic probe (Luxtron Corp., 1060 Terra

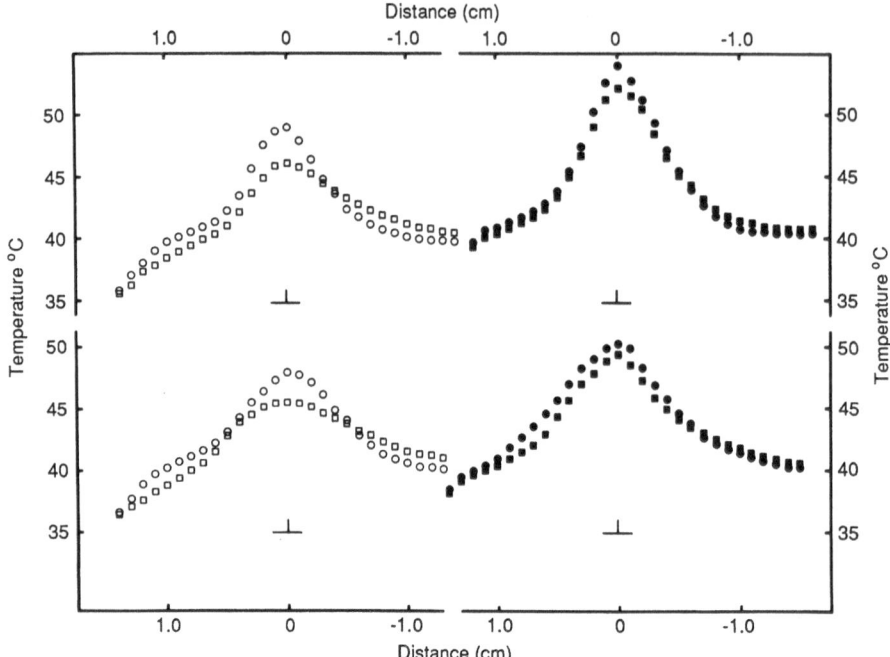

Fig. 1.28. Temperature profiles measured in normal canine tissue with two different probes in the vicinity of an MW interstitial heating antenna. Temperatures were measured within catheters (*open symbols*) and in bare tissue (*closed symbols*). The low thermally conducting optical probe (*circles*) measures significantly higher temperatures and a more narrow distribution than the high thermally conducting metal probe (*squares*). (Lyons et al. [61])

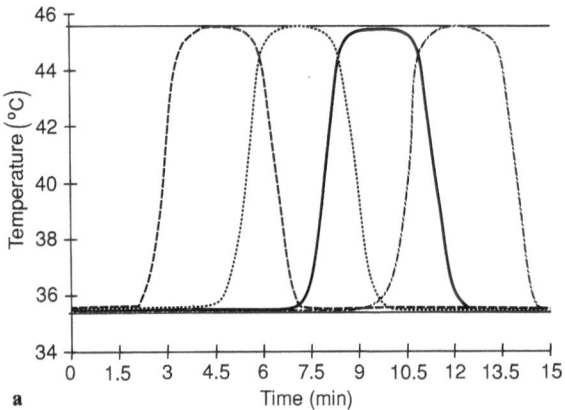

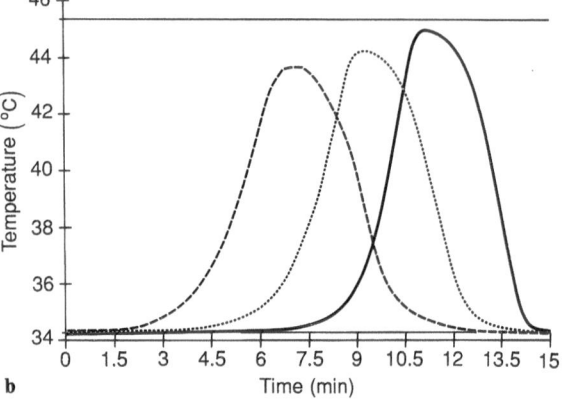

Fig. 1.29. Temperature measurements of a near step temperature function made with **a** a four-sensor fluoroptic probe and **b** a three-sensor metal thermocouple probe. The multisensor arrays are mechanically incremented inside a catheter through the step temperature distribution. The metal probe significantly distorts the profile owing to the thermal conduction along the metallic shaft. (Private communication, ER Lee, 1987)

Bella Ave., Mountain View, CA 94040) (Fig. 1.29a). The other is a three-sensor metal needle thermocouple probe (Sensortek, 154 Huron Ave., Clifton, NJ 07013) (Fig. 1.29b). In the presence of this rapidly changing temperature field, the differences in the measured profiles are quite striking. With the highly thermally conductive metal probe, the distortion or smearing of the step profile is obvious, with a notable depen-

dence on the sensor position within the metal needle shaft.

In such a controlled test, the probe-catheter configuration can be parameterized by a characteristic length (smearing or error length) that depends on the relative conductivity of the probe material, effective conductivity of the material surrounding the probe (e.g., tissue, catheter, air gap), and the probe dimensions [25, 62, 63]. The thermal smearing length, L, is associated with the functional form $\exp(-X/L)$, which is used to describe the response of a conductive temperature probe when used to measure a spatial step change in temperature. This parameter is analogous to the probe response time that characterizes the temporal response of a probe subjected to step change in temperature with time. Generally, the use of probes or probe-catheter combinations with large error lengths (≥ 1.5 mm) should be avoided in clinical treatments. Large gradients and the associated temperature field curvature are routinely observed at tissue-air or tissue-bolus interfaces, near interstitial heating electrodes, and at tumor-normal tissue boundaries. When sampling in such zones, the possibility of errors associated with thermal conduction should be minimized by using probe-catheter combinations that have thermal conductive properties similar to tissue, and by ensuring good thermal contact between the probe sensor, catheter, and tissue [61, 62].

A similar problem exists with respect to temporal smearing; that is, errors that arise due to the length of time required for a probe and sensor to achieve thermal equilibrium with the surrounding tissue. This lag or response time must be considered when transient data (e.g., rate of temperature rise for specific absorption rate measurements) are desired and during thermal mapping. The latter refers to the technique of moving a probe sensor mechanically through an indwelling tissue catheter [34, 35]. As the sensor is stepped along the catheter track, sufficient time at each measured point is required to ensure that temperature equilibrium has been established. The dwell time necessary to ensure the recommended temperature accuracy will depend on the profile being measured [35]. Response times for the probe-catheter combination need to be measured, and sufficient dwell time used to minimize the possibility of error when these mapping techniques are used [35, 65].

Finally, multisensor thermocouple arrays are very useful in clinical applications since they allow the efficient use of invasive probe tracks. However, such arrays are subject to error other than thermal smearing in high temperature gradients if they are constructed using the common wire technique. This technique uses a common constantan wire with several separated copper or manganin junctions soldered along this common constantan length [63]. The associated gradient error is the result of the thermoelectric emf-voltages generated within the finite dimensions of the junction region. In a very steep temperature gradient the junction region, because of its finite size, will not be isothermal. This effect is small for each isolated junction of reasonable dimension (e.g., ≤ 1 mm). However, the thermoelectric emf-voltages from several junctions (five to ten junctions) in a common wire type array are cumulative and can result in errors on the order of a degree in some clinical situations as one progresses from the first to last junction in the array. Thermocouple array construction techniques such as spot welding which minimize the junction size or use separate wire pairs for each junction will reduce or eliminate this source of error.

1.4.3 Electromagnetic Artifacts

The problems associated with EM-induced temperature artifacts have received considerable attention since MW and radiofrequency (RF) heating are the most common techniques [19, 66–71]. The presence of electrically conducting probe components allows preferential paths for direct or induced currents. The currents can result in selective heating of the probe, significant changes in the EM field pattern (Fig. 1.30), and EM interference with associated measurement equipment. The magnitudes of these undesirable effects are dependent on many factors; specifically, the effects are frequency dependent since capacitive and inductive coupling into the metal components play a role. There is also a strong dependence on the relative orientation of the electric field vector and the probe (Fig. 1.30). If the electric field is parallel to the conducting probe the resulting current will be maximum. Conversely, if the field is perpendicular to the probe, the current is minimal. EM interference with ancillary equipment can be caused by coupling through metallic probe leads or stray radiation leakage from heating applicators.

Thermocouple probes have particular problems in the EM environment since signal levels are small (microvolts to millivolts) and the different metals used may have significantly different electrical conductivities. The lead resistances, therefore, are not balanced and currents can readily flow through the junction, causing heating [19]. This is the case for the commonly used copper-constantan type junction. Special filters can be designed to balance the thermocouple leads or alternative metals can be used to reduce this effect [19, 25].

In addition to the probes, it has been shown that plastic catheters, particularly radiopaque catheters, have sufficiently different electrical properties from tissue that they can alter the EM field distribution [71]. This effect is also shown in Fig. 1.30.

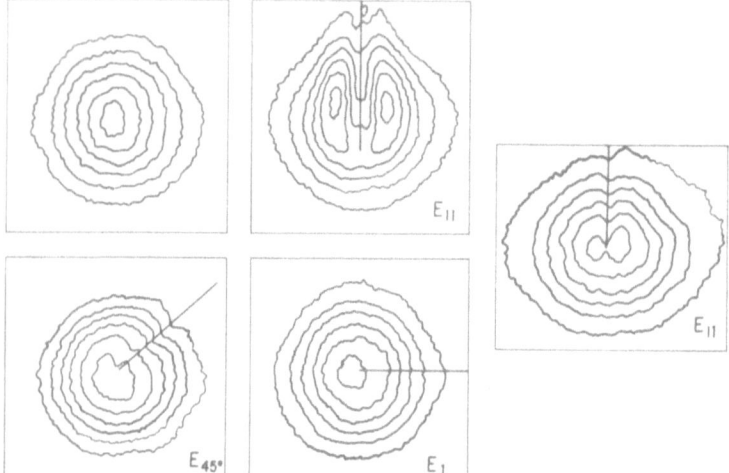

Fig. 1.30. Thermographic images of a plane within a tissue equivalent phantom taken after MW exposure with a heating applicator. The *upper left image* is the unperturbed distribution. The perturbation that results from a thermocouple probe inserted parallel ($E_{||}$), 45°, and perpendicular ($E_{\perp}$) to the electric field is sequentially shown in adjacent images. The perturbation of a radiopaque catheter parallel to the E-field is shown at the *right*. (Chan et al. [70, 71])

Several steps can be taken to minimize the problems associated with temperature measurement in EM-induced hyperthermia. The first and most effective solution is to use the available minimally perturbing EM probes and catheters. These technologies, however, are moderately expensive at present and until they become more cost-effective the standard metallic component probes (e.g., thermocouple, thermistors) will remain viable clinical options. If sensors with metallic components are to be used, the metals should be well insulated to prevent direct tissue-to-metal contact. Wire leads should be twisted to reduce magnetically induced currents, and the probe orientation should be chosen to be perpendicular to the electric field, if possible. Filters should be used to isolate the wire leads from measurement equipment. Power-off data sampling techniques should be used [19, 25].

1.4.4 Ultrasound Artifacts

Temperature measurements made during tissue heating via US remain problematic [64, 72–74]. Most plastic materials (e.g., biocompatible coatings, catheters) that are preferred for use in EM heating are not suitable for US heating. These materials generally have acoustic absorption coefficients several times higher than those of tissue [74]. Thus, they heat preferentially when directly exposed to US. Harder materials such as metals are less absorbent; however, the mismatch in acoustic properties between metal probes and tissues gives rise to viscous or frictional heating, which is caused by the relative motion of the metal probe and surrounding tissue [72]. Since the wavelengths of sound in tissue at useful therapeutic

US frequencies (0.3–3 MHz) are on the order of a millimeter (5.0–0.5 mm), the diameter of most practical implantable probes is a significant fraction of a wavelength. Thus, scattering of the US field by such probes can also be significant.

The combination of these effects (high absorption, viscous heating, scattering) introduces a level of uncertainty in US hyperthermia thermometry that can easily exceed the desired accuracy level of 0.1°–0.2°C. The best technique available at this time is to use the smallest practical bare metal needle probes (e.g., thermocouples and thermistors) directly implanted in tissue. These probes can still have viscous heating artifacts. Such artifacts can sometimes be identified by searching for sudden step-like changes in temperature when power is turned on or off (Fig. 1.31). When this occurs, it is possible to make approximate corrections for the preferential probe heating by subtracting this power-normalized step change from a specific probe's readings. As one might expect, this is a difficult correction to make, since the rise (or fall) time for the step temperature change is only a fraction of a second. This requires rapid data sampling for detection, and the correction generally has to be done retrospectively to treatment. The absorption artifact associated with plastics does not allow the use of indwelling catheters in US hyperthermia. This precludes the artifact-free thermal mapping that is commonly used in EM hyperthermia. Mechanically moving flexible thermocouple or fiber optic probes within a metallic needle track are an alternative that have found use in US treatments; however, thermal conduction smearing and US scattering become possible error sources in this approach. An ideal solution for invasive US thermometry is not yet available, and particular care is warranted in

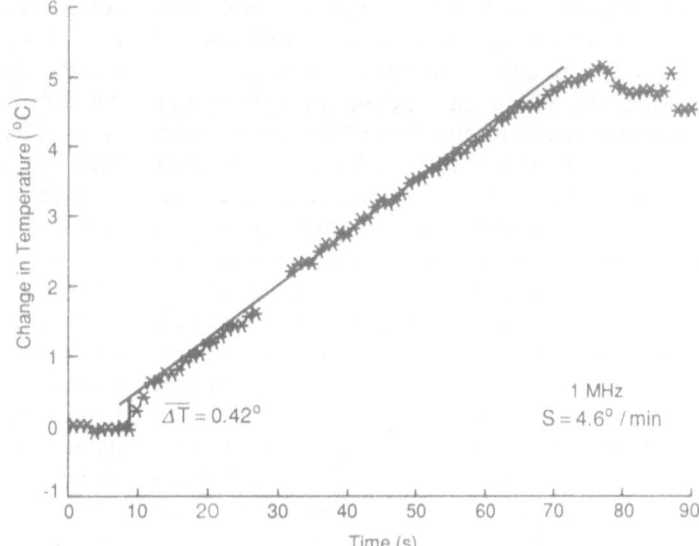

Fig. 1.31. A US-induced temperature artifact is demonstrated in an in vivo measurement. The artifact is characterized by a step-like rise in temperature ($\Delta T = 0.42\,°C$) when US power is applied at the 10-s mark. The temperature rise rate of the bulk medium is denoted by the slope indicated after the artifact heating (S = 4.6°/min)

clinical measurements. Temperature measurements can be made during short intervals when the US power is off in order to estimate temperature without artifact. Correcting these power-off measurements by extrapolating the temperature decay back to the time that power was terminated can improve the estimate of bulk tissue temperatures. In some cases, however, it may not be possible to decouple the viscous artifact from rapid cooling due to blood flow and thermal conduction. Also, the accuracy of the extrapolation correction will depend on the length of time the temperature decay curve is monitored. If one is using a scan-focused US heating technique, the scan path might be chosen in order to exclude the temperature probes from the focal zone of the US field, thereby minimizing artifact. This solution, however, necessarily precludes temperature measurement in the zone of high power absorption, which is not necessarily a desirable situation. Extra efforts should be made to quantify measurement error so that a meaningful comparison of treatments using US and EM techniques can be made.

1.5 Future Developments

Continued progress in hyperthermia therapy will require a more complete quantitative characterization of the thermal distributions obtained in clinical treatments. It is not likely that this can be accomplished using the invasive technology currently available. In a realistic scenario, only a few interstitial measure-

ment tracks can be utilized in most treatments. Since significant temperature heterogeneity is often observed, there are no, a priori, preferred sampling positions that can give an accurate representation of what tissue volume is heated to a given temperature. Therefore, better means for determining a more complete description of clinical temperature distributions are required. There are currently two alternatives. One is the development of noninvasive tomographic thermometry. A second is a combination of limited invasive or noninvasive thermometry with predictive modeling used for interpolation. Efforts are currently underway in both of these areas.

1.5.1 Noninvasive Thermometers

The functional requirements for noninvasive thermometry have been outlined by Cetas [75]. The most important requirement is to identify a measurable observable that depends on temperature, the temperature dependence of which can be isolated from other dependencies. Performance criteria for invasive thermometry, such as those listed in Table 1.1, should not be construed as absolute, since a relaxation of any one requirement may be advantageous in a given situation. For example, one might relax, by as much as an order or magnitude, the requirement for temperature or spatial resolution in order to obtain a more complete, although lower resolution, two- or three-dimensional noninvasive temperature mapping. Several methods have been proposed for noninvasive temperature measurement. These include magnetic resonance imaging (MRI), X-ray computed tomography (CT), ultrasonic computed tomography, MW and US radiometry, and applied potential tomography.

In MRI the relaxation mechanisms for nuclear magnetic moments involve thermal interactions with the surrounding medium. Two relaxation parameters are defined: the spin-lattice or longitudinal relaxation time T_1, and the spin-spin or transverse relaxation time T_2. Both of these characteristic relaxation times are temperature dependent; however, T_1 is probably the best suited for thermometry information, since it is generally a simple exponential relaxation process. Temperature coefficients for T_1 have been measured for some tissue types (e.g., muscle, blood, spleen) and these coefficients are on the order of 0.01 per °C. This requires that T_1 be determined to 1% in order to obtain 1 °C temperature resolution. It is estimated that, with current MRI technology, the time required for imaging with adequate signal averaging to obtain 1 °C precision will be of the order of several minutes [76, 77]. This time scale is too long to observe transient temperature changes that occur during heat-up and cooling, but may allow for measurement of a steady state condition. For the latter, however, the observations would have to be made concurrently with heating in order to maintain the steady state. Even though the constraints on signal levels and data acquisition times do not eliminate MRI as a possible approach to noninvasive thermometry, the clinical compatibility of MRI with current heating modalities (MW, RF, US) has yet to be demonstrated. In addition, the independence of the relaxation parameters on biological and physiological factors such as changes in blood flow or cellular biochemistry, which may be present in vivo, has to be verified. These should be the directions of future investigation.

X-ray computed tomography has also been proposed for noninvasive thermometry [78–80]. In this application the temperature-dependent parameter is the physical density of the material, which can be inferred by measuring linear attenuation coefficients for X-rays. In practice, one measures relative attenuation coefficients, where the reference value is water at a standard temperature. The density dependence of water and of tissue with high water content on temperature is relatively small [78–80]. When expressed in CT numbers or "Hounsfield units" (HU) the temperature coefficients are on the order of −0.4 HU/°C [79, 80]. Typically, the noise in a CT image, expressed as the standard deviation of CT numbers in an image of a uniform sample, is on the order of 5–10 HU. Therefore, one needs at least an order of magnitude gain in signal to noise to recover temperature-related changes in CT number. To some extent this can be ac-

complished by data-processing techniques such as averaging repeated CT scans, CT subtraction (i.e., subtracting the averaged CT numbers), and digital filtering techniques [79, 80]. Demonstrations of noninvasive CT thermometry in phantom and in vitro have been reported [79, 80]. In these controlled studies the temperatures obtained and spatial resolution achieved appear to be useful for hyperthermia thermometry (e.g., 0.25 °C at 2.5 cm^2 spatial resolution [79] and 0.5 °C at 0.42 cm^2 spatial resolution [80]). The technical problems associated with in vivo application are, however, substantial. The presence of motion artifact can easily defeat the image subtraction techniques used. Once again, the independence of CT number with regard to biology and physiology needs to be verified. Noninvasive CT thermometry does have desirable qualities regarding data acquisition time (scan times on the order of seconds) and is perhaps more suited than MRI to simultaneous use with present heating modalities.

Ultrasound tomography thermometry, with the temperature-dependent parameter being tissue attenuation or the speed of sound in tissue, has been investigated by several groups [28, 81–83]. Tomographic images of temperature distributions based on US absorption or speed changes in an in vitro phantom model consisting of beef tissue have been obtained [81]. Temperature and spatial resolution of the order of 0.5 °C and 1.0 cm^3 can be obtained with transit-time resolution of 1 ns and a signal to noise ratio of 10–15 [81]. In addition, the characterization of a number of tissue types with respect to the temperature dependence of US speed has been reported [82, 84]. One negative result from these tissue characterization studies is the observation that fat has a negative temperature coefficient where other examined tissues have a positive coefficient [84]. This anomaly leads to difficulties in tomographic image reconstruction in a heterogeneous sample. Other prominent difficulties associated with the thermometry concept include the sensitivity of US transmission and reflection at tissue interfaces, and the technical difficulties associated with reliably coupling the US beam(s) to the target volume in a clinical setting.

Radiometry in both the EM and US domains has been explicitly proposed for noninvasive thermometry [28, 51, 52, 85–89]. The principle involves detecting EM or US thermal noise radiation since the power spectrum of this noise can be associated with temperature via Planck's law. In the noninvasive setting a receiver (EM antenna or US transducer) is located outside the body and detects radiation originating in a subregion at depth. The size of the subregion that can be sampled is fundamentally limited by the wavelength of radiation. While this constraint is not a problem in US radiometry, it does severely limit EM radiometry. The latter limitation arises from the wavelength-dependent attenuation of thermal noise radiation as it passes through the absorbent tissue medium. Only the longer-wavelength portions of the EM noise spectrum can be detected with sufficient signal intensity from deep tissue sites; thus, spatial resolution is compromised in EM radiometry. In addition, for EM radiometry it is necessary to detect thermal radiation noise over a spectrum of frequencies in order to obtain unambiguous temperature information with respect to the depth in tissue. If the intensity of only one frequency (i.e., a very narrow band centered on one frequency) is detected, it is not possible to distinguish a warm region near the surface from a hotter region at a deeper depth, since the intensity of the signal from the deep hot region will be attenuated by intervening tissue. Thus, high gain EM radiometers capable of signal detection in several narrow frequency bands together with information about the interviewing tissue profile are necessary to reconstruct the temperature distribution as a function of depth.

In addition to the intrinsically shorter wavelength associated with US, another potential advantage of US radiometry is that temperature information from small tissue volumes at depth can be acquired by examining the time correlations of ultrasonic thermal noise signals in two or more detecting transducers [28, 29]. This is possible because the transient time of US emanating from a small volume deep in tissue and detected at the body surface can be resolved with present technology. Therefore, cross-correlating, as a function of delay time, the US noise received by two transducers effectively samples temperature information from a specific region of interest at depth in tissue. Although US radiometry based on this concept has been proposed, it has not yet been technically developed. Therefore, its potential advantage for use in hyperthermia with regard to sampling at depth and spatial resolution has not been demonstrated.

Microwave radiometry has been put into practice and multiband radiometers have been tested by several groups [85, 86, 89]. In one case, this thermometry approach has been incorporated into a clinical system and is the principal means of temperature measurement and control for this superficial MW heating system [52]. Nonetheless, the degree to which these radiometric technologies can replace or supplement invasive temperature measurements needs to be demonstrated. Again, there exists the complication of interference between the radiometric measurement and means for heat induction. The possibility that these functions can be time shared using the same US

transducer or EM antenna is one approach to the solution of this problem [85, 86].

Applied potential tomography has also been suggested as a noninvasive temperature measurement concept [90, 91]. Using this technique, a two-dimensional image of tissue dielectric properties can be constructed from complex impedance measurements made at several locations on the skin surface [92]. Generally, the complex dielectric properties (dielectric constant and conductivity) of tissues are temperature dependent. Therefore, an image subtraction technique similar to that mentioned for the other tomographic imaging technologies might be used to extract a two- or three-dimensional temperature distribution.

Applied potential imaging is now in the early developmental stages, so its feasibility for thermometry is in the process of being established [93]. It has the advantage that the necessary surface impedance measurements may be done in a physically simple way, without the constraints of associated bulky hardware in close proximity to the patient. Thus, this technique may be more compatible with the various other hyperthermia hardware needed for heating.

1.5.2 Mathematical Modeling

An alternative and perhaps more feasible method for characterizing thermal distributions is to invoke the physics of bioheat transfer as a basis for calculating the temperature distributions. Efforts is this area are being aggressively pursued [30, 31, 94–97]. The problem is complicated, but amenable to numerical calculation techniques using modern computer resources.

The first step in this process is the modeling of energy deposition in tissue that results from external or interstitial heating sources. This means solving (probably in three dimensions for realistic clinical situations) an appropriate EM or US field equation in heterogeneous media. Results using finite difference and finite element numerical techniques have been reported [94]; however, the majority of these cases are limited to EM fields in two-dimensional simulations.

Assuming that a satisfactory solution for the energy deposition problem is obtained, the second step is to incorporate this source term into a model for bioheat transfer, again in three-dimensional heterogeneous tissue media. In this solution, energy transfer by thermal conduction and blood flow must be modeled accurately. Heat transfer via thermal conduction is well understood and is formulated in various numerical methods. The required thermal parameters are the thermal conductivity and diffusivity for various tissue types, and have been, or can be accumulated [98]. Thus, the solution to this aspect of the heat transfer problem presents no fundamental difficulty. In contrast, the role of blood flow in bioheat transfer is not well understood. It is, however, clear that blood flow heat transfer is a substantial contributor to the energy balance equation and is perhaps responsible for much of the temperature heterogeneity observed in clinical treatments. The situation is complicated by the fact that blood flow during hyperthermia changes as a function of both time and space [99]. Most present models for blood flow heat transfer are handled mathematically in the simplistic form of a time invariant heat sink [100]. In some applications the strength of the sink term is used as an adjustable parameter, which can be varied to optimize the agreement between measured and predicted temperature distributions [95–97]. The problem of blood flow heat transfer remains a fundamental challenge in the mathematical modeling of in vivo temperature distributions [101, 102].

The results of mathematical modeling with regard to predicting observed temperature distributions have not been rigorously tested. The obvious questions associated with spatial dimensionality, blood flow heat transfer, accurate specifications of boundary conditions, and numerical and measurement errors are areas for future research.

1.6 Summary

Invasive thermometry technologies that meet most of the performance criteria needed for safe clinical applications of hyperthermia treatments are available. However, the invasive measurement approaches limit the acquisition of temperature data, making it difficult to characterize and quantify individual treatments. Multipoint measurements made using either mechanical mapping or multisensor probes should be routinely carried out in all treatments, in order to take full advantage of the few interstitial paths that are practical in routine clinical applications. Although these data may be insufficient to definitely answer important therapeutic questions concerning thermal dose or treatment optimization, these invasive temperature measurements give insight into how well

selected tissue volumes are heated, and provide direction for improvements in both temperature measurement and heating equipment.

Several thermometry concepts have been developed into clinically usable systems with temperature probes adequate for invasive temperature monitoring during hyperthermia. These systems offer a reasonable choice to today's hyperthermia practitioner. A summary of the advantages (+) and disadvantages (−) of these available technologies is given in Table 1.2. The more traditional systems based on thermocouples and thermistors have the general advantages of being affordable and reliable. However, they have the disadvantage of being susceptible to artifact when used in conjunction with the intense US or EM fields used for tissue heating. Adapting these systems for the simultaneous use with heating equipment increases the cost and requires attention to ensure that these adaptive techniques are properly employed.

The optical thermometry concepts represent relatively new technologies and, as such, are less cost-effective than the traditional systems. Their specific advantage rests in the fact that the plastic or glass fiberoptic probes have low thermal conductivity and do not couple to EM fields. However, since they are made of plastic materials, the probes are subject to absorption artifacts in US fields. The probes themselves have the potential for being a low cost, even disposable, item. However, this realization requires further technical development and expanded applications in other fields since therapeutic hyperthermia is not a large enough commercial application to support cost-efficient mass production.

Hyperthermia therapy in the future will surely demand more in the way of invasive temperature measurement in order to allow active feedback control of therapy and make use of mathematical algorithms for predicting temperature distribution based on these limited measured data. Noninvasive thermometry remains to be demonstrated clinically feasible. Except for microwave radiometry, most of the proposed noninvasive thermometry concepts are far removed from clinical applications and would appear to require substantial efforts in technical development. On the other hand, mathematical modeling of in vivo temperature distributions does not depend as much on new technological advances, but does require accurate formulations of complex physiological heat transfer processes, computational codes for identifying three-dimensional tissue heterogeneity, definition of appropriate discrete volume elements, and efficient numerical algorithms for generating mathematical solutions for power deposition and temperature fields.

Table 1.2. Advantages and disadvantages of various thermometry systems

Thermocouple thermometry

Advantages (+)

1. Intrinsic calibration
2. Linear response (single point calibration)
3. Size: multisensor array probes ≤1 mm o.d.
4. Multichannels (16+) systems
5. Cost-effective

Disadvantages (−)

1. Small signal levels
2. High thermal conduction
3. EM interference and artifact
4. US viscous artifact

Thermistor thermometry

Advantages (+)

1. High accuracy and resolution
2. Good long-term stability
3. Size: multisensor array probes <1 mm o.d. when used with metallic leads
4. Adapted for EM fields with high-resistance leads
5. Multichannel (16) systems

Disadvantages (−)

1. Not intrinsically calibrated
2. Nonlinear response
3. High thermal conduction when used with metallic leads
4. EM interference and artifact with metallic leads
5. US viscous artifact
6. Limited to single sensor probes when high-resistance leads are used

GaAs optical thermometry

Advantages (+)

1. EM immunity
2. Low thermal conductivity
3. Size: single sensor <0.5 mm o.d. multisensor ≤2 mm o.d.
4. Multichannel (12) system

Disadvantages (−)

1. Not intrinsically calibrated
2. Nonlinear response
3. Bend artifact
4. US absorption artifact

Fluoroptic (spectrum ratio) thermometry

Advantages (+)

1. EM immunity
2. Linear response (two-point calibration)
3. Low thermal conduction
4. Size (single sensor 0.8 mm o.d.)

Disadvantages (−)

1. Not intrinsically calibrated
2. Performance problems (drift − lamp life)
3. Bend artifact
4. Not available in multisensor array
5. US absorption artifact
6. Limited to a four-channel system

Fluoroptic (decay time) thermometry

Advantages (+)

1. EM immunity
2. Linear response

Disadvantages (−)

1. Intrinsic calibration limited to ±2 °C

(single-point calibration)
3. Low thermal conductivity
4. Size: multisensor array ≤1.0 mm o.d.
5. Multichannel (8) system

2. US absorption artifact

Acknowledgment. The authors wish to acknowledge and thank Jeanne Forest and Gerard Honoré for their assistance in preparing this manuscript.

References

1. Hahn G (1982) Hyperthermia and cancer. Plenum, New York
2. Overgaard J (1987) Some problems related to the clinical use of thermal isoeffect doses. Int J Hyperthermia 3:329–336
3. Kapp DS, Fessenden P, Samulski TV, Bagshaw MA, Cox RS, Lee ER, Lohrback AW, Meyer JL, Prionas SD (1988) Phase I evaluation of equipment for hyperthermia treatment of cancer. Int J Hyperthermia 4:75–115
4. Overgaard J (1985) Rationale and problems in the design of clinical studies. In: Overgaard J (ed) Hyperthermic oncology, vol 2. Taylor and Francis, London, pp 325–338
5. Oleson JR, Sim DA, Manning MR (1984) Analysis of prognostic variables in hyperthermia treatment of 163 patients. Int J Oncol Biol Phys 10:2231–2239
6. Arcangeli G, Arcangeli G, Guerra A, Lovisolo G, Cividallie A, Marino C, Mauro F (1985) Tumor response to heat and radiation: prognostic variables in the treatment of neck node metastases from head and neck cancer. Int J Hyperthermia 1:207–217
7. van der Zee J, van Putten WLJ, van den Berg AP, van Rhoon GC, Wike-Hooley JL, Broekmeyer-Reurink MP, Reinhold HS (1986) Retrospective analysis of the response of tumours in patients treated with combination of radiotherapy and hyperthermia. Int J Hyperthermia 2:337–349
8. Dunlop PRC, Hand JW, Dickinson RJ, Field SB (1986) An assessment of local hyperthermia in clinical practice. Int J Hyperthermia 2:39–50
9. Kapp DS, Samulski TV, Fessenden PF, Bagshaw MA, Lee ER, Lohrbach AW, Cox RS (1987) Prognostic significance of tumor volume on response following local-regional hyperthermia and radiation therapy. 7th Annual Meeting NAHG, February 1987, Atlanta GA
10. Rozzell TC, Johnson CC, Durney DH, Lords JL, Olsen RG (1974) A nonperturbing temperature sensor for measurements in electromagnetic fields. J Microwave Power 9:241
11. Bowman RR (1976) A probe for measuring temperature in radiofrequency heated material. IEEE Trans Microwave Theory Tech MTT-24:43–45
12. Cetas TC (1976) Temperature measurement in microwave diathermy fields: principles and probes. In: Robinson JE, Wizenberg MJ (eds) Proceedings of the international sym-

posium on Cancer Therapy by hyperthermia, drugs and radiation. ACR, Bethesda, p 193

13. Christensen DA (1977) A new non-perturbing temperature probe using semiconductor band edge shift. J Bioeng 1:541−545

14. Cetas TC, Connor WG (1978) Thermometry considerations in localized hyperthermia. Med Phys 5:79−91

15. Wickersheim KA, Alves RB (1979) Recent advances in optical temperature measurement. Ind Res/Dev 21:82−89

16. Samulski TV, Shrivastava PN (1980) Photoluminescent thermometer probes: temperature measurements in microwave fields. Science 208:193−194

17. Sholes RR, Small JG (1980) Fluorescent decay thermometer with biological applications. Rev Sci Instrum 51:882

18. Barth P, Angell T (1982) Thin linear thermometer arrays for use in localized cancer hyperthermia. IEEE Trans on Electron Devices ED-29 (1):144−150

19. Chakraborty DP, Brezovich IA (1982) Error sources affecting thermocouple thermometry in RF electromagnetic fields. J Microwave Power 17:17−28

20. Olsen RG, Hammer WC, Taylor JC (1982) Coaxial nonmetallic thermocouple with electronic ice point for dosimetric use in electromagnetic environments. J Microwave Power 17 (2):137−143

21. Samulski TV, Chopping PT, Haas B (1982) Photoluminescent thermometry based on erropium-activated calcium sulphide. Phys Med Biol 27 (13):107−114

22. Vaguine VA, Christensen DA, Lindley JH, Walston TE (1984) Multiple sensor optical thermometry system for application in clinical hyperthermia. IEEE Trans Biomed Eng BME-31:168−172

23. Wickersheim KA (1986) A new fiberoptic thermometry system for use in medical hyperthermia. SPIE Proceedings, vol 713

24. ter Haar GR, Dunn F (1984) Linear thermocouple arrays for in vivo observation of ultrasonic hyperthermia. Br J Radiol 57:257−258

25. Carnochan P, Dickinson RJ, Joiner MC (1986) The practical use of thermocouples for temperature measurement in clinical hyperthermia. Int J Hyperthermia 2:1−19

26. Cetas T (1982) Invasive thermometry. In: Nussbaum GH (ed) Physical aspects of hyperthermia. American Institute of Physics, New York, pp 231−265

27. Hand JW (1985) Thermometry in hyperthermia. In: Overgaard J (ed) Hyperthermic oncology, vol 2. Taylor and Francis, London, pp 299−308

28. Christensen D (1982) Current techniques for noninvasive thermometry. In: Nussbaum GH (ed) Physical aspects of hyperthermia, American Institute of Physics, New York, pp 266−270

29. Cetas TC (1987) Thermometry. In: Field SB, Franconi C (eds) Proc NATO advanced study institute on physics and technology of hyperthermia, Urbino. Nijhoff, Dordrecht, pp 470−508

30. Roemer RB (1987) Thermal models/temperature distribution. In: Field SB, Franconi C (eds) Proc NATO advanced study institute on physics and technology of hyperthermia, Urbino. Nijhoff, Dordrecht, pp 553−561

31. Roemer RB (1987) Treatment planning and evaluation. In: Field SB, Franconi C (eds) Proc NATO advanced study institute on physics and technology of hyperthermia, Urbino. Nijhoff, Dordrecht, pp 562−573

32. Hahn GM (1987) Non-invasive methods of temperature measurement. In: Field SB, Franconi C (eds) Proc NATO advanced study institute on physics and technology of hyperthermia, Urbino. Nijhoff, Dordrecht, pp 509−516

33. Shrivastava P, Luk KH, Oleson J, Dewhirst M, Pajak T, Paliwal B, Perez C, Sapareto S, Saylor T, Steeves R (in press, 1989) Hyperthermia Quality Assurance Guidelines. Int J Radiat Oncol Biol Phys

34. Gibbs FA (1983) "Thermal mapping" in experimental cancer treatment with hyperthermia: description and use of a semiautomatic system. Int J Radiat Oncol Biol Phys 9:1057−1060

35. Engler MJ, Dewhirst MW, Winget JM, Oleson JR (1987) Automated temperature scanning for hyperthermia treatment monitoring. Int J Radiat Oncol Biol Phys 13:1377−1382

36. Luk KH, Pajak TF, Perez CA, Johnson RJ, Comner N, Dobbins T (1984) Prognostic factors for tumor response after hyperthermia and radiation. In: Overgaard J (ed) Hyperthermic oncology, vol 1. Taylor and Francis, London, pp 353−356

37. Dewhirst MW, Sim DA (1986) Estimation of therapeutic gain in clinical trials involving hyperthermia and radiotherapy. Int J Hyperthermia 2:165−178

38. Samulski TV, Fessenden PF, Valdagni R, Kapp DS (1987) Correlations of thermal washout rate, steady state temperatures and tissue type in deep seated recurrent or metastatic tumors. Int J Radiat Oncol Biol Phys 13:907−916

39. van der Zee J, van Rhoon G, Broekmeijen-Reurink M, Reinhold H (1987) The use of implanted closed-tip catheters for introduction of thermometry probes during local hyperthermia treatment series. Int J Hyperthermia 3:337−345

40. Sapozink MD, Gibbs FA, Gates KS, Stewart JR (1984) Regional hyperthermia in the treatment of clinically advanced, deep-seated malignancy: results of a pilot study employing an annular array applicator. Int J Radiat Oncol Biol Phys 10:775−786

41. Bagshaw MA, Taylor MA, Kapp DS, Meyer JL, Samulski TV, Lee ER, Fessenden PJ (1984) Anatomical site-specific modalities for hyperthermia. Cancer Res [Suppl] 44:4842s−4852s

42. Knundsen M, Heinzl L (1986) Two-point control of temperature profile in tissue. Int J Hyperthermia 2:21−38

43. Lele PP (1982) Local hyperthermia by ultrasound. In: Nussbaum GH (ed) Physical aspects of hyperthermia. American Institute of Physics, New York

44. Hynynen K, Roemer RB, Anhalt D, Johnson C, Moros E (1985) Focussed, scanned ultrasound for local hyperthermia. IEEE Proc 7th Annual Conf of IEEE Eng in Med Biol Soc, pp 341−345

45. Ogilvie GK, Goss SA, Badger CW, Burdette EC (1986) Performance of a multi-sector ultrasound hyperthermia applicator and control system: results of animal studies in vivo. 34th Annual Meeting of Radiation Research Society, Atlanta

46. Strohbehn JW, Mechling JA (1986) Interstitial techniques for clinical hyperthermia. In: Hand JW, James JR (eds) Physical techniques in clinical hyperthermia. Research Studies Press, Letchworth, pp 210−287

47. Fessenden P, Kapp DS, Lee ER, Samulski TV (1988) Clinical microwave applicator design. In: Paliwal BR, Hetzel FW, Dewhirst MW (eds) Biological, physical and clinical aspects of hyperthermia. Med Phys Monogr 16:123−131

48. Turner PF (1986) Interstitial EM applicator/temperature

probe. IEEE Proc 8th Annual Conf of the IEEE Eng in Med and Biol Soc, vol 3, pp 1454–1457

49. Prionas SD, Goffinet DR, Samulski TV, Fessenden P, Hahn GM (1984) Characterization of an interstitial hyperthermia RF system utilizing flexible electrodes. 32nd Annual Meeting of the Rad Res Soc, 1984, Orlando

50. Sun MH, Samulski TV (1987) Catheters with integrated fiberoptic temperature sensors for monitoring and control of interstitial hyperthermia. 7th Annual Meeting of NAHG, February 1987, Atlanta

51. Paglione RW, Sterzer F, Wozniak FJ (1986) Noninvasive thermometry for a robot-controlled microwave hyperthermia system. IEEE Proc 8th Annual Conf of the IEEE Eng in Med and Biol Soc, vol 3, pp 1500–1502

52. Plancot M, Prevost B, Fabre JJ, Chive M, Moschetto Y, Giaua G (1986) Thermal dosimetry based on radiometry in multi-media. IEEE Proc 8th Annual Conf of the IEEE Eng in Med and Biol Soc, vol 3, pp 1429–1434

53. Thermocouple Reference Tables (1979) NBS Monograph 125, National Bureau of Standards, Washington, DC

54. Shrivastava P, Saylor T, Matloubieh A, Paliwal B (1988) Hyperthermia thermometry evaluation: criteria and guidelines. Int J Radiat Oncol Biol Phys 14:327–335

55. Steinhardt JS, Hart JR (1968) Calibration curves for thermistors. Deep Sea Res 15:497–503

56. Ryan TP, Britt RH (1984) Computer controlled temperature multiplexer for thermal dosimetry studies. J Biomed Eng 6:302–304

57. Sostman HE (1977) Melting point of gallium as a temperature calibration standard. Rev Sci Inst 48:127–130

58. Mangum BW, Thornton DD (1977) NBS Spec Pub 481. US Government Printing Office, Washington DC

59. Mangum BW (1983) Triple point of succinonitrile and its use in the calibration of thermistor thermometers. Rev Sci Inst 54:1687–1692

60. Wise JA, Soulen Jr RJ (1986) "Thermometer calibration", a model for state calibration laboratories. (NBS Monograph 174) US Government Printing Office, Washington DC

61. Lyons BE, Samulski TV, Britt RH (1985) Temperature measurements in high thermal gradients. I. The effects of conduction. Int J Radiat Oncol Biol Phys 11:951–962

62. Samulski TV, Lyons BE, Britt RH (1985) Temperature measurements in high thermal gradients: II. Analysis of conduction effects. Int J Radiat Oncol Biol Phys 11:963–971

63. Dickinson RJ (1985) Thermal conduction errors of manganin-constantan thermocouple arrays. Phys Med Biol 30:445–453

64. Fessenden P, Samulski TV, Lee ER (1984) Direct temperature measurement. Cancer Res [Suppl] 44:4799s–4804s

65. Waterman FM (1985) The response of thermometer probes inserted into catheters. Med Phys 12:368–372

66. Johnson CC, Guy AW (1972) Nonionizing electromagnetic wave effects in biological material and systems. Proc IEEE 60:692–718

67. Samaras GM, Rosenbloom S, Cheung AY (1983) Correction of microwave-induced thermistor sensors errors. Med Phys 10 (3):326–332

68. Dunscombe PB, McLellan J, Malaker K (1986) Heat production in microwave-irradiated thermocouple. Med Phys 13:457–461

69. Constable RT, Dunscombe P, Tsoukatos A, Malaker K (1987) Perturbation of the temperature distribution in microwave irradiated tissue due to the presence of metallic thermometers. Med Phys 14:385–388

70. Chan KW, Chou CK, McDougall JA, Luk KH (1988) Changes in heating patterns due to perturbations by thermometer probes at 915 and 434 HMz. Int J Hyperthermia 4:447–456

71. Chan KW, Chou CK, McDougall JA, Luk KH (1988) Perturbations due to the use of catheters with non-perturbing thermometry probes. Int J Hyperthermia 4:699–702

72. Fry WJ, Fry RB (1954) Determination of absolute sound levels and accoustic absorption coefficients by thermocouple probes-theory. J Acoust Soc Am 26:294–310

73. Hynynen K, Martin CJ, Watmough DJ, Mallard JR (1983) Short communications. Errors in temperature measurement by thermocouple probes during ultrasound induced hyperthermia. Br J Radiol 56:969–970

74. Kuhn PK, Christensen DA (1986) Influence of temperature probe sheathing material during ultrasound heating. IEEE Trans Bio Med Eng BME-33:536–538

75. Cetas TC (1984) Will thermometric tomography become practical for hyperthermia treatment monitoring. Cancer Res (Suppl) 44:4805s–4808s

76. Parker DL, Smith V, Sheldon P, Crooks LE, Fussell L (1983) Temperature distribution measurements in two-dimensional NMR imaging. Med Phys 10 (3):321–325

77. Parker D (1984) Applications of NMR imaging in hyperthermia: an evaluation of the potential for localized tissue heating and noninvasive temperature monitoring. IEEE Trans Biomed Eng BME-31:161–167

78. Zemenhof RG, Sternick ES, Curran BM (1981) Non-invasive temperature mapping by computerized tomography. Int J Radiat Biol Phys 7:1235

79. Fullone B, Moran P, Podgorsak E (1982) Noninvasive thermometry with a clinical X-ray CT scanner. Med Phys 9 (5):715–721

80. Bentzen SM, Overgaard J, Jorgensen J (1984) Isothermal mapping in hyperthermia using subtraction X-ray computer tomography. Radiother Oncol 2:255–260

81. Johnson S, Christensen D, Baxter B, Greenleaf J, Rajagopalan B (1977) Noninvasive acoustic temperature tomography for measurement of microwave and ultrasound-induced hyperthermia. J Bioenerg 1:555–570

82. Bowen T, Conner WG, Nasoni RL, Piter AE, Sholes RR (1977) Measurement of the temperature dependence of the velocity of ultrasound in soft tissues. 2nd Int Symp on Ultrasonic Tissue Characterization, Gaithersburg

83. Nasoni RL (1982) In vivo temperature dependence of the speed of sound in mammalian tissue and its possible use in hyperthermia. Proc Third Int Symp Cancer Therapy by Hyperthermia, Drugs and Radiation. NCI Monogr 81:501–504

84. Bamber J, Hill C (1979) Ultrasonic attenuation and propagation speed in mammalian tissue as a function of temperature. Ultrasound Med Biol 5:149–157

85. Robilland M, Chive M, Leroy Y, Audet J, Pichot C, Bolomey J (1982) Microwave thermography – characteristics of waveguide applicators and signatures of thermal structures. J Microwave Power 17 (2):97–105

86. Sterzer F, Paglione R, Wonziak F, Mendecki T, Fridenthal E, Botstein C (1982) A self-balancing microwave radiometer for non-invasively measuring the temperature of subcutaneous tissue during localized hyperthermia treatments of cancer. IEEE Microwave Theory Techniques MTT-S Digest, vol 1, 438–440

87. Mizushina S, Oh-ishi H, Hamamura Y (1984) A three-band microwave radiometer for noninvasive temperature measurement. IEEE Microwave Theory Techniques MTT-S Digest, vol 1, 145–147

88. Prionas S, Hahn G (1985) Noninvasive thermometry using multiple-frequency-band radiometry: a feasibility study. Bioelectromagnetics 6:391–404

89. Mizushina S, Hamamura Y, Sugiura T (1986) A three-band microwave radiometer system for noninvasive measurement of temperature at various depths. IEEE Microwave Theory Techniques MTT-S International Microwave Symp, Baltimore MD

90. Conway J, Hawley MS, Seagar AD, Brown BH, Barber DC (1985) Non-invasive thermal monitoring by an applied potential tomographic method (APT), Strahlentherapie 161:428

91. Griffiths H, Antal J, Ahmed A (1985) Non-invasive thermometry using applied potential tomography. Strahlentherapie 161:534

92. Barber D, Brown B (1984) Applied potential tomography. J Phys E: Sci Instrum 17:723–733

93. Bach Andersen J (1987) Quasi-states conductivity reconstruction of a cylinder. Int J Hyperthermia 3:548 (abstract)

94. Strohbehn J, Roemer R (1984) A survey of computer simulations of hyperthermia treatments. IEEE Trans Biomed Eng BME-31:136–149

95. Divrik A, Roemer R, Cetas T (1984) Inference of complete tissue temperature fields from a few measured temperatures: an unconstrained optimization method. IEEE Trans Biomed Eng BME-31:150–160

96. Paulsen KD, Strohbehn JW, Lynch DR (1984) Theoretical temperature distributions produced by an annular phased array-type system in CT-based patient models. Radiat Res 100 (3):536–552

97. Dewhirst M, Winget J, Edelstein-Keshet L, Sylvester J, Engler M, Thrall D, Page R, Oleson J (1987) Clinical applications of thermal isoeffect dose. Int J Hyperthermia 3:307–318

98. Bowman HF (1981) Heat transfer and thermal dosimetry. J Microwave Power 16 (2):121–133

99. Song CW (1984) Effect of local hyperthermia on blood flow and microenvironment: a review. Cancer Res (Suppl) 44:4721 s–4730 s

100. Pennes HH (1948) Analysis of tissue and arterial temperature in the resting human forearm. J Appl Physiol 1 (2):93–122

101. Chen MM, Holmes KR (1980) Microvascular contributions in tissue heat transfer. In: Jain RK, Gulino PM (eds) Thermal characteristics of tumors: applications in detection and treatment. YY Acad Sci 335:137–150

102. Weinbaum S, Jiji LM (1985) A new simplified bioheat equation for the effect of blood flow on local average tissue temperature. J Biomech Eng Trans ASME 107:131–139

103. Hand JW, Lagendijk JJW, Bach Anderson J, Bolomey JC (1989) Quality assurance guidelines for ESHO protocols. Int J Hyperthermia 5:421–428

104. Dewhirst M, Phillips T, Samulski T, Stauffer P, Shrivastava P, Paliwal B, Sapozink M, Myerson R, Leeper D, Fessenden P, Kapp D, Oleson J, Emami B (1990) RTOG Quality assurance guidelines for clinical trials using hyperthermia. Int J Rad Onc Biol Phys (in press)

2 Noninvasive Control of Hyperthermia

J.C. BOLOMEY and M.S. HAWLEY

2.1 Introduction

It is now well demonstrated that hyperthermia can constitute an efficient adjuvant to radiotherapy and chemotherapy in cancer treatment. The precondition of successful results is that hyperthermia treatments are carefully controlled according to more or less well-defined clinical protocols stipulating the most efficient treatment sequences. These protocols imply some constraints concerning the temperature distribution tolerances to be satisfied in the heated volumes in order simultaneously to obtain therapeutic efficiency inside tumoral tissues and to avoid any undesirable burning effect in surrounding healthy tissues.

The possibility of such temperature control is of prime clinical importance for the adjustment of the heating equipment. In this respect, interstitial hyperthermia provides an efficient approach, when possible, because of (a) the spatial localization of the treated volume between implanted electrodes and (b) the available access for temperature measurements inside the electrodes. Similarly, superficial treatments allow suitable atraumatic temperature controls. It is probably not pure coincidence that the most significant results of hyperthermia, used alone or as an adjuvant, have been obtained with interstitial and superficial modalities, demonstrating the crucial importance of the temperature control.

Such a conclusion is reinforced when one observes the remaining difficulties affecting cases of deep or semideep and/or large tumoral volumes. The complexity of these cases mainly results from unfavorable biological and physical processes and is greatly aggravated when trying to achieve heat deposition by means of external noninvasive heating equipment through electromagnetic or ultrasonic waves. Usually, temperature controls are invasively performed by using thermocouples or optical fibers. As a result, the temperature can be considered to be known with good accuracy, but only at a finite number of discrete points or continuously along a segment. Outside the points of measurement, it is only possible approx-imately to appraise the amplitude of the temperature gradients, which, in some cases, has been estimated to be a few degrees per centimeter, for instance in the neighborhood of bones. Lack of elevation of the temperature at the points of measurement does not exclude the possibility that the temperature is dangerously increased at unexpected hot points, provoking pain and/or damage without any therapeutic efficiency. It is worth noting that the invasiveness of classical thermometric means may constitute a limit to the acceptance of the hyperthermia treatment by the patient. Sometimes, the hyperthermia session is conducted without any internal temperature control, because the patient would not tolerate it.

The first salient feature of present deep treatments is, then, that temperature control can be done at an insufficient number of points and that the traumatic aspect of classical thermometry constitutes a serious handicap for effective evaluation of the effectiveness of hyperthermia treatments. The second is that, with most of the existing equipment, there is an evident lack of flexibility to better match the temperature constraints imposed by the clinical protocols. The available parameters are the operating frequency, the shape and the location of the electrodes or applicators, and finally the power. For a given arrangement, the operator can modify the power delivered by the generator and try to do so in such a way that this power is presumably dissipated in the considered volume by cancelling the so-called reflected power by means of a suitable matching unit. These adjustments are sometimes performed automatically from recorded temperature values at some test points that are significant or are expected to be so. In more sophisticated equipment, including multielectrodes and applicators, some individual adjustment of each of them is offered as a facility to the operator in order to better control the power deposition according to complex and time-varying thermoregulatory processes. But such an increase in the available degree of freedom requires additional data and criteria in order to optimize the adjustments with a view to obtaining the desired effect. Furthermore, initial adjustments are needed before delivering the nominal power level. Finally, the use of multiparameter equipment which is, a priori, expected to be the only way to improve the power deposition in a given volume, implies broader spatial control of this volume and of its environment, before and during the hyperthermia treatment.

As a general consequence, it clearly appears that, beside the strict temperature aspect of hyperthermia control, there is a real need for an extension of the control concept. This need is confirmed by clinical experience demonstrating that unexplained factors limit the efficiency of hyperthermia treatments. These factors are, for instance, responsible for the variability of the results obtained with the same patient from one session to another, even when the same protocol is used. As observed effects, the temperature at the measured points cannot reach the therapeutic efficiency level even if maximum power is used, or this level cannot be reached because the pain felt by the patient prevents any power elevation. The reasons for such a variability may have different causes. Indeed, the combination of electrodes/applicators, including boluses, with the patient constitutes a very complex load — in the electrical sense of the world — the behavior of which with respect to the generator can be modified by several technical or physiological factors. For instance, in capacitive hyperthermia, the electrical impedance of the load seen by the generator, and, consequently, the power transfer between the generator and its load, may depend on the salinity of the water in the boluses or on the electrode arrangement but also on the electrical characteristics of the biological tissues located between the electrodes. If, in the first case, appropriate technical controls can compensate for what can be considered as parasitic effects, in the second case the variations of the electrical properties of tissues may unavoidably result from the modification of these tissues under the influence of associated complementary therapies. The complexity of the phenomena involved during hyperthermia treatments should then lead to consideration of what systematic and noninvasive controls should be achieved in order to guarantee the reproducibility of the conditions under which the treatment will be delivered.

All these factors have generated a real interest in noninvasive control techniques and stimulated research efforts in that direction. Since treatment efficiency has until now almost exclusively been considered in terms of temperature levels, most of these efforts have been devoted to noninvasive thermometry. As a matter of fact, analysis of the published literature reveals two main directions of investigation. The first concerns the improvement of already existing techniques which are based on infrared or microwave radiometry. Such radiometric techniques are indeed the only ones to be effectively used in clinical situations and have proven their practical usefulness, at least in the case of rather superficial tumors. Consequently, it appears quite natural to try to extend their range of applicability to deeper configurations.

The second direction of investigation aims to take advantage of different imaging modalities through a convenient decoding of the implicit temperature content available in images related to the interaction of tissues with electromagnetic or acoustic waves. The rapid and spectacular development of tomographic techniques (X-ray tomodensitometry, NMR imaging, ultrasound echotomography, etc.) has performed an important role in such approaches, on which are based the greatest expectations. However, these approaches are much more complicated than one could imagine, as demonstrated by the fact that not one technique has emerged as significantly preferable from among all the possible imaging modalities which have been investigated. Periodically, pertinent and well documented papers [24, 28, 56, 81] present the current state of the art and attempt to appraise the respective merits of different approaches.

The imaging techniques which have been taken into account in the last few years, and the progress of which is regularly reported, are radiometry (infrared, microwave, or acoustic), X-ray tomodensitometry, NMR imaging, various ultrasound modalities, and applied potential tomography. More recently, microwave tomography and thermo-induced ultrasonic emission have been considered. Laser transillumination is a still newer technique. But comparison between these techniques is very difficult for three main reasons which are now briefly analyzed.

First, the imaging modalities under consideration may be at very different levels of development. Some of them have already been operative for a few years, and their use is rapidly increasing. In contrast, other approaches are still at the conceptual level, being more or less supported by theoretical and/or numerical modelization. Between these two extremes, some imaging modalities are just starting to come into use and only a few prototypes are available globally for preliminary experiments. Such differences render comparative study very difficult.

Secondly, the claimed performances either do not correspond to perfectly comparable configurations or result from doubtful extrapolations. This fact mainly results from the diversity of the techniques involved in an imaging system. Quite often, the limitations stemming from basic physical phenomena and those deriving from the equipment are not distinctly separated, with the result that it is very difficult to decide whether the actual performances are or are not perfectible. Even when some possibility of improvement exists, it is important to evaluate the required cost because economic factors have a decisive impact on the practical usefulness of a given technique.

Finally, the imaging techniques are developed in different physical and technical settings between which the flow of information is limited because of pure mutual ignorance or for other reasons related to competition or industrial property. Such a situation does not facilitate objective comparisons.

The purpose of this chapter is to try to offer some comprehensive data for objective comparison of the performances obtained, or to be obtained at reasonable cost, with some imaging modalities using electromagnetic or ultrasonic waves. For this purpose the physical basis of these modalities will be introduced in order to illustrate, as well as possible, their fundamental limitations in simple and comparable configurations. In addition, some results obtained with the more realistic configurations will also be presented. The complexity of the task means that exhaustive coverage of this very large and changing field of investigation is out of the question.

This chapter is organized into seven chief sections. Section 2.2 is devoted to general considerations on the basic approaches to noninvasive thermal control. The required short- or long-term performances of different control modalities are briefly indicated. Different imaging approaches to noninvasive thermal control are classified according to several aspects of fundamental and practical importance. The problems associated with the interpretation of tomographic images are discussed with regard to the particular objective of thermal sensing. Quality criteria of images are then introduced, leading to the concept of performance optimization, e.g., temperature resolution vs spatial resolution, of an imaging process under different constraints. More particularly, the general problems associated with the use of an imaging technique for noninvasive thermal control during hyperthermia treatments are considered.

Section 2.3 deals with electromagnetic radiometric techniques. The microwave case is considered from a prospective point of view (Chap. 3 in this volume is concerned with the present state of the art in microwave radiometry). Fundamental limitations are discussed and recent developments of existing techniques are indicated, with particular reference to imaging ability.

Sections 2.4 and 2.5 concern X-ray tomodensitometry and NMR imaging, respectively. As these imaging techniques are among the most popular, their principles are only very briefly reviewed, and most of each section is devoted to examples extracted from the existing literature.

Section 2.6 probably constitutes the most original part of this chapter, being related to dielectric imaging modalities. These approaches present some salient differences from other better known techniques, especially with regard to image quality criteria. While usual optical criteria apply to most popular techniques, dielectric images require a greater interpretative effort in order to profit from their content. Such differences must be pointed out as clearly as possible to avoid too hurried comparisons or exaggerated expectations. However, the basic short- and long-term potential of these dielectric imaging techniques − and, more particularly, their sensitivity − justifies the importance attributed to them in this chapter, even if one can reasonably consider that many improvements and further developments remain to be achieved. The authors hope that their own experience in impedance tomography, microwave imaging, and inverse scattering has not excessively unbalanced the content of the chapter in favor of what they know the best but is perhaps generally less well-known in the medical domain.

Section 2.7 is devoted to acoustic approaches. In many respects, some active tomographic procedures are similar to those described in the previous section, provided acoustic properties of tissues are substituted for dielectric ones. Passive radiometry is also addressed, as is thermo-induced acoustic imaging.

Finally, Section 2.8 attempts to synthesize all the available results and to point out the most promising solutions when all the aspects of the problem are taken into account. It is shown that no universal solution exists to noninvasive thermometry leading to general purpose equipment. But, as compared to the actual situation where invasive means are almost exclusively used, some noninvasive techniques should already be able to provide useful information in specific configurations.

Considering so many different technical domains has made it impossible to adopt a completely coherent list of symbols throughout the seven sections. At the risk that the same symbol can represent different quantities in different sections, the deliberate choice has been to favor the most used notation in the area under consideration. Any resulting ambiguity should be rapidly removed by the lists of symbols given after formulas and, more generally, by the context.

Most of the formulas are given without detailed derivations, which may be found easily in the appropriate references. Formulas have been limited in number, simply providing comprehensive support in the understanding of the influence of parameters of interest. Similarly, some paragraphs may seem shorter than their importance merits, but the reason is that the same subject is detailed in another chapter or volume of this series.

Finally, the authors are expressing their own opinions in the light of their present state of the art knowledge; this knowledge derives from the published literature and from many kind contributions by, or discussions with, specialists of different areas.

2.2 General Considerations Regarding Imaging Technique Performances

2.2.1 Presentation

This first section is devoted to general purpose considerations on noninvasive control of hyperthermia, on possible goals in view of clinical practice, and on imaging techniques themselves. The need for general comments on imaging techniques and an attempt at classification results from their diversity, from existing a priori ideas or misunderstanding of real technical problems involved in imagery, or, more simply, from the partitioning between different and sometimes concurrent approaches. Even if the natural trend is to extend image quality criteria of usual optical instrumentation to all of these approaches, the same concepts (spatial resolution, contrast, etc.) must be carefully handled because they may have very different implications and significance from a practical point of view.

A very general formalism is used in order to illustrate the specific difficulties when using different imaging techniques to achieve noninvasive control of hyperthermia treatments. Particular attention is devoted to showing how the performances of a given technique result from a necessary compromise between opposing fundamental and practical constraints.

2.2.2 Expected Performances for Noninvasive Control of Hyperthermia

As already explained in the introduction, the term "noninvasive control of hyperthermia" covers several aspects. First of all, since according to some analogy with radiotherapy, therapeutic efficiency is, at least

possibly, related to the thermal dose concept, *noninvasive thermometry* constitutes the goal of prime evident importance. Reliable knowledge of the thermal dose needs the temperature to be accurately measured within a sufficient volume and with an adequate spatial resolution. From this point of view, and leaving aside various considerations of a practical nature to be detailed in later sections, the performances of noninvasive controls can be technically summarized as follows:

1. Temperature sensitivity/stability/accuracy of the order of $1\,°C$
2. Spatial resolution of a few millimeters
3. Response time of a few seconds

However, the authors are convinced that, between these ultimate goals and the present situation, there is much room for more modest, but nevertheless interesting, performances.

Noninvasive thermometry implicitly means that the result of the measurement depends only on the temperature at the measurement point and on nothing else. The problem is not so simple in the hyperthermia context even with classical invasive probes (thermometers, thermocouples, optical fibers, etc.), which can be sensitive, for instance, to the heating radiation or to inconvenient thermal contacts. However, it can be reasonably assumed that the measurement is independent of the local tissue properties and only derives from some specific behavior of the probe (apparent volume of an expandable liquid, electromotive force at the contact between two different metals, etc.).

The situation becomes much more critical when the temperature measurement is achieved via the thermal behavior of a characteristic of the tissue under test by itself. For the considered characteristic, the temperature sensitivity is obviously dependent on the tissue composition, which can be altered by the hyperthermic treatment. Blood flow rate changes, edema formation, and necrosis are among the most significant alterations. How temperature measurements are affected by such alterations depends on the selected noninvasive technique. Their real influence is actually often not yet very well known, but it is being seriously investigated more and more.

Coming back to the noninvasive control of hyperthermia, the lack of absolute thermometry does not necessarily mean that very useful therapeutic information cannot be obtained from the observation of change in the blood flow rate, for instance. Such indirect effects of temperature changes could be more sensitive than first order effects. For evident reasons, it is very difficult to give accurate figures to quantify the expected performances in the measurement of such quantities.

Finally, it is worth noting that hyperthermic treatment efficiency is strongly, but not uniquely, dependent on the power deposition capability of the arrangements of electrodes or applicators to be used. The last few years have demonstrated that the first enthusiastic expectations of producing power deposition by means of external devices were far too optimistic. Even appropriate selection of the operating frequency and of the applicator geometry and positioning does not provide a convenient solution to the difficult cases of deep and/or large tumoral volumes. The combination of arrays of applicators appears to represent a possible improvement. But, as already mentioned, increasing the number of available parameters does not necessarily facilitate the task of the clinicians, unless noninvasive means of controlling the SAR distribution in the desired area are available. Beside real clinical situations, preliminary studies on phantoms constitute an important phase of investigation which must not be underestimated because it usually involves invasive and long probing procedures.

As a first conclusion, it is clear that several noninvasive control concepts have to be considered and that this justifies looking at the contribution of the available imaging and tomographic techniques.

2.2.3 Classification of Imaging Modalities

The objects are remotely discerned by means of the waves they radiate. Images of these objects can be formed by using a suitable analog or numerical treatment of their radiations measured at convenient locations. Existing imaging modalities can be classified according to different and complementary aspects such as:

1. Fundamental physical mechanism of radiation
2. Nature of the radiated waves
3. Frequency range, or wavelength to dimension ratio, or interaction mode
4. Experimental procedures
5. Numerical algorithms of tomographic reconstruction
6. Present level of development

A first classification can be done according to the physical mechanism from which the radiation results (Fig. 2.1). With *passive imaging modalities*, the object

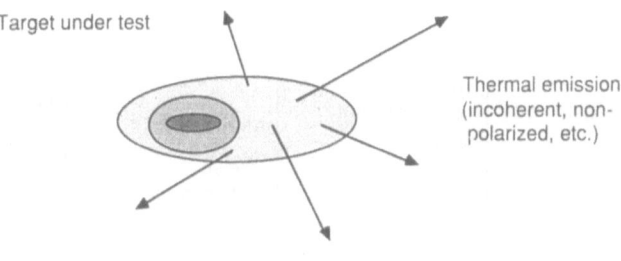

Passive imaging

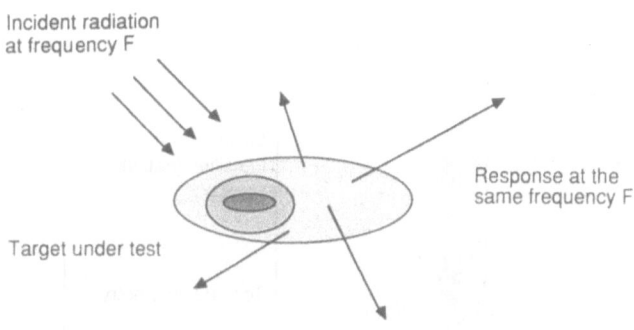

Fig. 2.1. Passive and active imaging configurations. In the passive mode, the object can be seen by itself, from the incoherent wave it radiates. In the active mode, the object is visible from its response (reflection, refraction, diffraction, etc.) to a given illuminating wave whose characteristics constitute additional available parameters

Active imaging

is assumed to radiate spontaneously, within a broad frequency spectrum, in the absence of any particular illumination. The relevant techniques are called radiometric or thermographic because this radiation originates in thermal noise emission produced by charge motions or mechanical deformations occurring in any object at a given temperature. In contrast, *active modalities* require the object to be illuminated by an incident radiation. The object is then seen from its response to this incident radiation. The mechanisms underlying this response may be more or less complicated, involving scattering or diffraction, for instance. In general, the linearity of the response guarantees that the frequency of the response is the same as that of the incident radiation. A third class of imaging modalities concerns *mixed mechanisms*: an external stimulation is used to increase the natural thermal noise emission. It is worth noting that mixed approaches differ from passive ones in that in the lat-

ter case that external stimulation is natural and not provoked. All these imaging modalities have been effectively used for biomedical purposes but are presently at very different levels of development.

A second classification of all the possible imaging modalities can be based on the nature of the radiation. In this chapter, the objects will be considered on the basis of their ability to interact with *electromagnetic* or *acoustic* waves. Both these waves can be used in passive, active, or mixed imaging, as defined before. As shown in Fig. 2.2, a very large part of the spectrum of electromagnetic waves has been used in imaging techniques. This spectrum itself can be roughly divided into two parts with regard to the ionizing capabilities of the corresponding radiation. The upper part of the electromagnetic spectrum corresponds to *ionizing radiations*, among which X-rays have been extensively used for medical imaging for many years. Their wavelengths are much smaller than

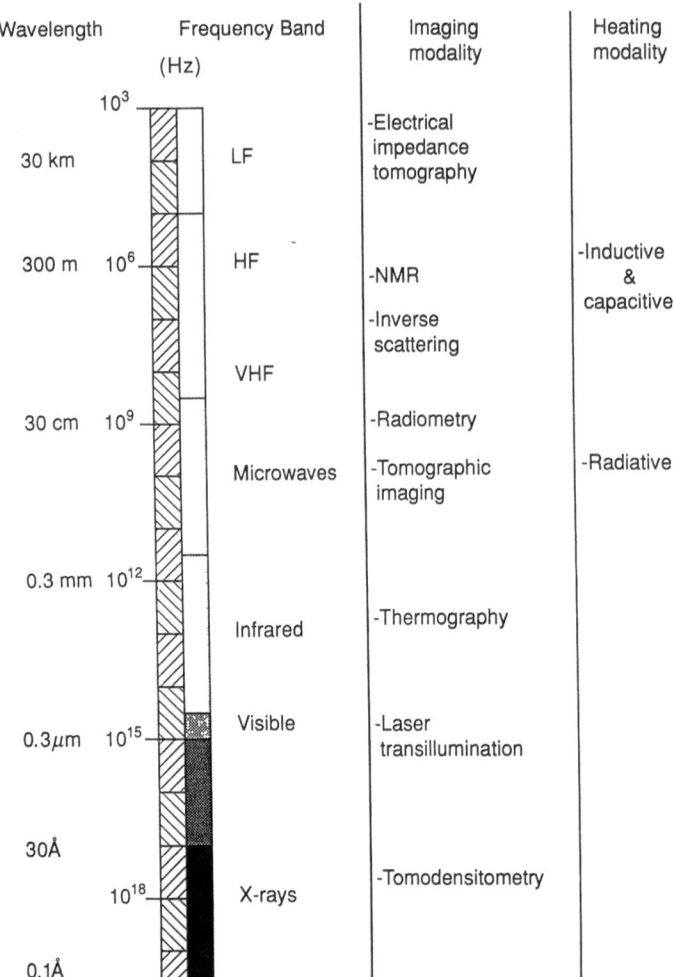

Fig. 2.2. Different imaging and heating modalities over the whole electromagnetic wave spectrum

in the lower part of the spectrum. Due to their energy, X-rays propagate in living tissues according to straight linear trajectories. Such a simple situation added to the possibility of obtaining narrow and well-collimated ray beams has significantly helped the development of the now popular tomodensitometry.

In a broad lower part of the electromagnetic spectrum extending from low frequencies to the visible optical range, *no ionizing effects* can be observed. More precisely, this frequency range covers from a few kilohertz to about 10 THz. There is thus a ratio of the order of 10^{12} between the extreme frequencies. In terms of wavelengths in free space, the variation extends from 300 km to 0.3 µm. That is to say that, according to the interval of the spectrum under consideration the wavelength can be much greater or smaller than the observed biological targets, even if one takes into account that the wavelength in high water content tissues is only a fraction of the free space wavelength.

The mechanism of interaction between waves and tissues is strongly dependent on the ratio between the wavelength and the dimensions of the body or of the detail under observation. For the same organ, for instance, the electromagnetic radiation may interact very differently. Visible light is scattered by muscle tissues, microwaves are diffracted by bones or tissue discontinuities, and low-frequency (LF) or high-frequency (HF) currents circulate within human bodies according to classical circuit theory.

At low frequencies, quasistatic approximations allow the use of circuit concepts such as voltage, currents, or impedances. At these frequencies, a complex body can be represented by means of its electrical equivalent scheme. In contrast, for very high frequencies these concepts become meaningless and are replaced by an optical description of the phenomena. For this purpose, ray tracing techniques have proved very convenient.

Whereas visible light mainly propagates along linear or slightly refracted paths, microwaves as well as LF and HF currents do not propagate so simply. Even in homogeneous media, they cannot be easily concentrated or collimated. Nevertheless, in all cases the interaction is governed by the electromagnetic (electrical or optical) properties of tissues at the considered frequency. These electrical properties, consisting of dielectric constant and conductivity − or the complex index in optical terminology − are themselves very variable with frequency. Furthermore, they more or less directly depend on other physical or physiological factors such as water content, blood flow rate, and temperature.

An important difference between X-rays and waves of the lower part of the electromagnetic spectrum is that the former are incoherent and unpolarized whereas the latter are generally coherent and polarized, even in the visible range thanks to lasers. While X-rays produce intensity images only, the other waves with larger wavelengths can be used to produce different kinds of image according to the displayed quantity. Amplitude and phase can be utilized as well. Furthermore, the vector nature of electromagnetic fields allows consideration of the polarization aspect to exhibit copolarization or cross-polarization effects.

The case of ultrasonic waves looks quite different mainly because the part of the ultrasound spectrum used in medical applications has been, until now, much narrower. This spectrum extends from about 100 kHz to a few megahertz, corresponding to wavelengths in living tissues in the range of 1 cm to a fraction of a millimeter. There is only a factor of 10^3 between extreme values of frequencies, instead of 10^{12} for electromagnetic waves. The general propagation regime is thus usually governed by diffraction laws. Tissues can be characterized from their sound velocity and absorption.

As already stated, the general interaction properties of a wave, whether electromagnetic or ultrasonic, with some structures of given dimension strongly depend on the wavelength to dimension ratio. Consequently the same analysis can be conducted, at least formally, in both cases if the wavelength to dimension ratio is comparable. Nevertheless, the result of the interaction can be very different, depending on the acoustic and electromagnetic characteristics of the tissues. As is well known, there is some complementarity between the behavior of the same tissue with respect to electromagnetic and acoustic waves. Generally speaking, a given tissue is able to propagate preferably one kind of these waves.

Finally, imaging techniques can be roughly classified according to the newest evolution of the image concept. The first images on the retina or on the plate of a photo camera can be considered as reflected or transmitted projections of the observed scene on the recording surface. Even this simple process contains the germs of the posterior tomographic technique. Indeed, the projection is considered as satisfactory only if the body under investigation lies between some distance limits, e.g., the punctum proximum and the punctum remotum in the case of the eye. A convenient modification of these distances can usually be obtained by means of some adjustment of the imaging system, providing de facto some tomographic capabilities, even with a single view-angle procedure. This capability is particularly interesting for optical systems with large numerical aperture and small field depth such as microscopes which allow the observa-

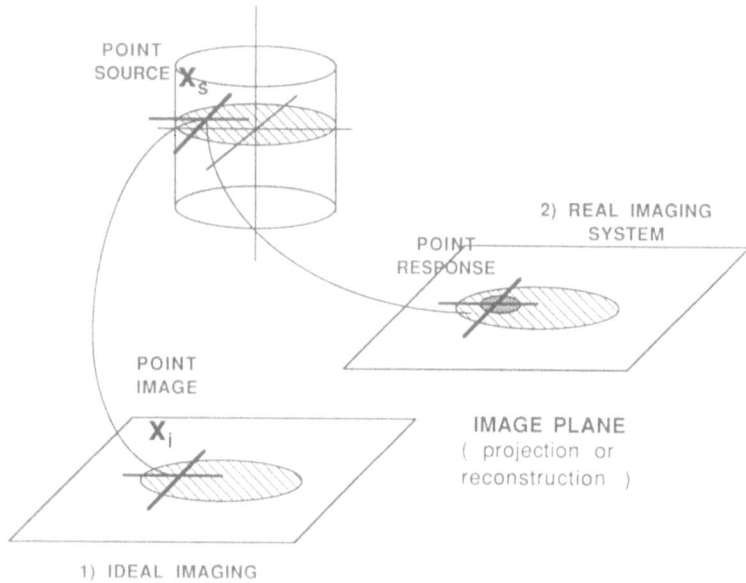

Fig. 2.3. Comparison between ideal and real imaging systems as regards their spatial resolution. An ideal system provides a point to point correspondence between the object and its image. With a real system, the response of a point object is a small volume in the image space (point response)

tion of small objects slice by slice, with a thickness of the order of a wavelength. The recent advent of multi view-angle procedures has evidently improved the situation with regard to tomographic reconstructions.

2.2.4 Analysis of Image Content

The analysis of the content of an image depends on its *quantitative* or *qualitative* aspect. In order to illustrate the difference between these two aspects, the ideal case of a perfect tomographic system will be considered (Fig. 2.3). Such a system would provide a point to point correspondence between a source point x_s and an image point x_i. For active or mixed modalities, the image intensity at a point x_i can be written, when omitting a proportionality coefficient, as follows:

$$I(x_i) = Q\{x_s, p_1(x_s), p_2(x_s), \ldots p_N(x_s)\} \cdot I_0(x_s) \qquad (2.1)$$

or, more briefly as:

$$I(x_i) = Q(x_s, p_s) \cdot I_0(x_s) \; ; \quad \text{with } p_s = p(x_s) \qquad (2.2)$$

where Q is a quantity which is directly related to the electromagnetic or acoustic properties of the observed medium, and I_0 is a weighting function depending on the illumination of this medium. Q generally varies with the observation point − or, in other words, with the nature of the tissue − and with different physical or physiological factors p_n (n = 1, 2 ... N) at the same point. The final image intensity I results from digital and/or analog processing of measured data.

As examples, Table 2.1 gives different quantities accessible by means of different imaging modalities.

In quantitative imaging, the illumination can be considered as constant or uniform, in such a way that the image intensity I is directly proportional to the quantity Q. By means of a suitable calibration, quantitative measurement of Q can be achieved. Such an approach, when possible, can be used for tissue characterization purposes. Conversely, if the considered medium is homogeneous, i.e., Q is constant, then the image provides the distribution of I_0 in the explored volume.

For qualitative imaging, the image intensity depends not only on the intrinsic properties of the tissues but

Table 2.1. Some quantities and the related imaging modalities

Quantity (Q)	Imaging modality	Wave
Density Proton density	X-ray tomodensitometry	EM
Relaxation times (T_1, T_2) Chemical shift	RF, NMR imaging	EM
Acoustic properties (absorption, velocity, etc)	Echotomography, TOF (Time of flight) measurements Velocimetry	US
Electrical properties (conductivity, complex permittivity)	Electrical impedance tomography	EM
	Microwave tomography	EM

also on the way in which they are illuminated. Such a situation is typical in the case of multiple scattering phenomena between waves and tissues. In a homogeneous medium, the illumination function I_0 represents the illumination wave distribution. In quantitative imaging, this function is independent of Q and/or known. But, more generally, I_0 depends on Q. Introducing the general relationship existing between I_0 and Q:

$$I_0(\mathbf{x_i}) = H\{Q(\mathbf{x_s}, \mathbf{p_s})\} , \qquad (2.3)$$

the image intensity appears to be nonlinearly related to Q by the following equation:

$$I(\mathbf{x_i}) = Q(\mathbf{x_s}, \mathbf{p_s}) \cdot H\{Q(\mathbf{x_s}, \mathbf{p_s})\} . \qquad (2.4)$$

Despite its apparent complexity, qualitative imaging can be very useful. As a convincing example, let us just remember that our optical perception of the environment belongs to this imaging approach. Indeed, we know that the image we have from an object depends on its lighting. In the limiting case of darkness, the objects cannot be seen! On the other hand, the same object can produce several images according to the illumination angle while its optical properties remain unchanged. It is evident that our previous experience and our brain data-processing capabilities play an important role in the object recognition from partial and or different views, or in presence of parasitic images created by some reflecting interface or refracting layers.

The possibility of quantitative imaging constitutes a significant advantage for further interpretation of resulting images. Qualitative imaging requires some appropriate training to decode the image content. As a matter of fact, the possiblity or the impossibility of quantitative imaging depends on the way in which images can be formed from measured data. This in turn depends on the complexity of the interaction mechanisms between tissues and waves. From a general point of view, and as far as such a point of view concerns tomographic reconstructions, the only problems to be efficiently numerically solved are linear problems. Efficiency is here understood in terms of rapidity and accuracy.

If, usually, the image intensity is obtained via a linear inversion problem from measured data, the reconstruction of the quantity $Q(\mathbf{x_s})$ requires the influence of the illumination function to be removed. Sometimes, this can be achieved either exactly or approximately according to the considered modality. Although not impossible, the inversion of the nonlinear equation to derive $Q(\mathbf{x_s})$ from $I(\mathbf{x_i})$ poses difficult problems associated with the influence of noise corruption of the measured data.

2.2.5 Usual Criteria of Image Quality

Usual criteria of image quality are naturally derived from our intuitive evaluation of optical images. They mainly consist of *spatial resolution* and *contrast*. Spatial resolution is a measurement of the ability of the imaging equipment to discern between two neighboring points. This quality has been introduced in the vision of very small apparent objects by means of microscopes or telescopes. More generally, spatial resolution is essential to relate the image intensity $I(\mathbf{x_i})$ to a more or less localized part of the object under observation. An important concept is the *point response* $P(\mathbf{x}; \mathbf{x}')$, which is the image intensity at $\mathbf{x}$ obtained from an ideal point source whose image should be located at $\mathbf{x}' = \mathbf{x_i}$ with a perfect instrument. The point response usually extends over a small volume centered around $\mathbf{x}'$ (Figs. 2.3, 2.4). The dimensions of this volume allow the definition of the minimum distance between two points in order that they are resolved by the instrument.

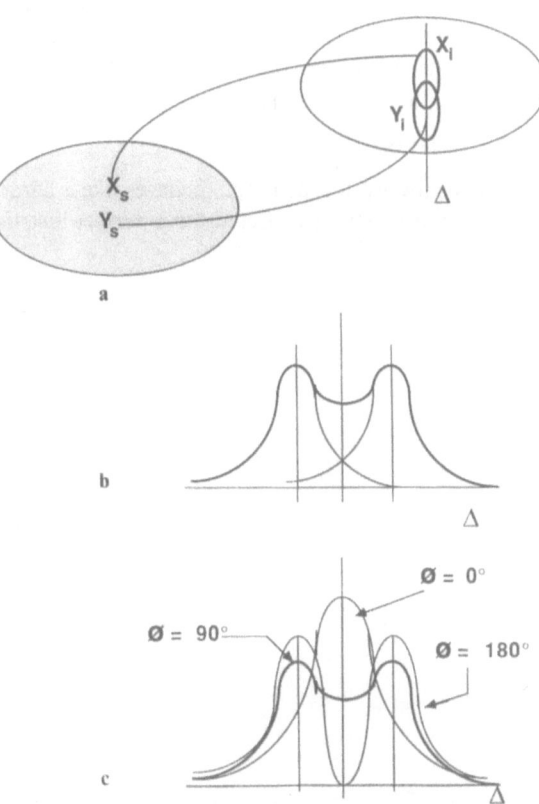

Fig. 2.4a–c. Limitation of the resolving power by finite spatial resolution. **a** Basic situation for the correspondence between two objects and their images. **b** Incoherent case; image intensity along the Δ line. **c** Coherent case; influence of the phase of the emitting point sources; image intensities along the Δ line

For optical, or quasioptical, instruments, the resolution is limited by diffraction phenomena. Such a limitation is also involved in all cases where the wavelength is of the order of the dimensions of the structure to be imaged or of the size of the illuminating devices. Roughly speaking, the point response extends over distances of the order of the wavelength. It is well known that, according to the Rayleigh criterion, the spatial resolution is a fraction of the wavelength. Two distinct points will be resolved if their distance Δx is larger than the half value width W of the point response P. Figure 2.4 illustrates schematically the resolution problem between two closely spaced points. It also points out the difference between the coherent and the incoherent cases for two small emitting source objects. In the incoherent case, the intensities (or powers) of the corresponding point responses add together, and the resolution of the two objects at a given wavelength mainly depends on their distance. In the coherent case, the field add together and the relative phase performs an important role for a given distance, leading to some possible confusion in the image interpretation.

More precisely, in the optical case, the effective intensity of the image provided by the instrument usually can be written as the convolution product of the ideal image I by the point response P:

$$I_{eff}(\mathbf{x}) = I(\mathbf{x}) * P(\mathbf{x};\mathbf{X}') . \tag{2.5}$$

$I_{eff}(\mathbf{x})$ tends toward $I(\mathbf{x})$ as $P(\mathbf{x};\mathbf{x}')$ behaves like a Dirac distribution $\partial(\mathbf{x}-\mathbf{x}')$, as expected for a perfect instru-

ment. The spatial extent of P tends to smooth possible abrupt variations of the displayed quantity Q. Furthermore, the image size W_i does not reflect the dimensions of the object W_o as far as it is smaller than W_{pr} (Fig. 2.5). Once again, the coherent and the incoherent cases present different situations. In the formation of the image of an edge, the coherent case will introduce "parasitic" fringes parallel to the edge direction while the incoherent case will exhibit a smooth transition between both sides of the edge.

In the limiting cases where the wavelength is much larger or smaller than the target dimensions, factors other than diffraction limit the spatial resolution, which is no longer directly related to the wavelength. Sampling rates, sensor sizes, and electrical and mechanical stabilities are some factors that can limit the spatial resolution. These factors concern both the measuring equipment and the numerical data processing which follows. As typical examples, electrical impedance tomography − or applied potential tomography −, which operates in the 100-kHz range, corresponding to a wavelength of 1 km, provides a spatial resolution of the order of 1 cm; on the other hand, spatial resolution of approximately a few tenths of a millimeter can be obtained by means of X-ray tomodensitometry, for a wavelength of the order of 1 nm. Nevertheless, the concept of point response still remains very useful for the determination of the spatial resolution.

It is worth noting that in any case the spatial resolution is ultimately limited by the equipment performances and more particularly by its sensitivity and dynamic range. Indeed, it can be shown that, from a purely theoretical point of view, even in so-called diffraction-limited situations the spatial resolution could be rendered as small as desired by means of a suitable increase in the ability of the measurement chain to process data varying between very small and very large values. Practically, however, it is quite impossible to statisfy such requirements with existing equipment when interaction mechanisms are governed by diffraction phenomena.

Spatial resolution is not the only way to appreciate the potential of an imaging modality. The contrast has to be introduced in order to establish whether it will be possible to distinguish some details from their respective intensity. While the spatial resolution concerns the localization of details, the contrast pertains to their visibility. The contrast is mainly governed by the local tissue behavior with respect to the interacting wave, but, as shown later, the total noise of the measurement chain has to be taken into account for a complete discussion. In other words, two different parts of an object, or the modification of one part of

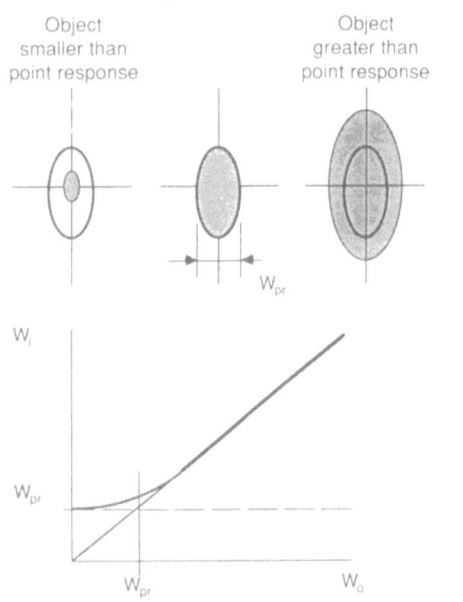

Fig. 2.5. Correspondence between object and image sizes due to finite spatial resolution

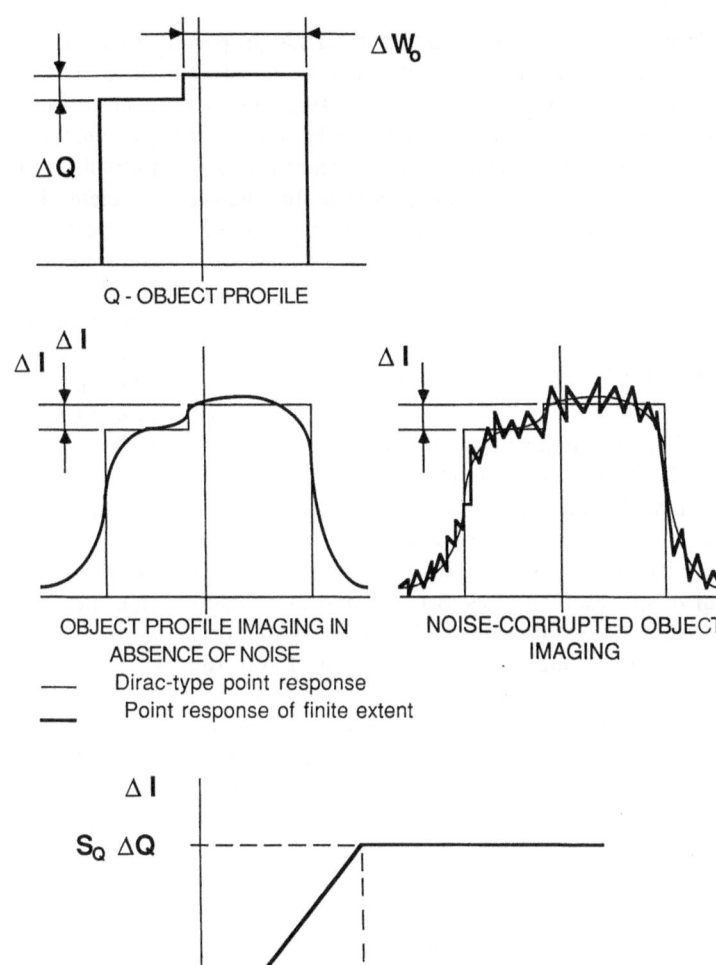

ΔW_o

ΔQ

Q - OBJECT PROFILE

ΔI ΔI

OBJECT PROFILE IMAGING IN
ABSENCE OF NOISE
— Dirac-type point response
— Point response of finite extent

ΔI

NOISE-CORRUPTED OBJECT
IMAGING

ΔI

$S_Q \Delta Q$

N

O W_{pr} ΔW_o

Fig. 2.6. Combined influence of noise and limited spatial resolution on image size and intensity

it, will not be visible, even with a perfect instrument, if these parts, or the same part before and after its modification, do not interact differently with the incident wave.

The contrast can be understood as the sensitivity of images with respect to the displayed quantity $Q(\mathbf{x}, \mathbf{p})$ and, consequently, to the different factors $\mathbf{p}(\mathbf{x})$ which $Q(\mathbf{x})$ is dependent upon. Mathematically, the contrast can be defined as the total derivative with respect to Q, or a partial derivative with respect to a given factor p_i:

$$S_Q(\mathbf{x_0}) = \left(\frac{dI}{dQ}\right)_{x = x0} , \qquad (2.6)$$

$$s_p(\mathbf{x_0}) = \left(\frac{\partial I}{\partial p}\right)_{x = x0} . \qquad (2.7)$$

Now, from a practical point of view, a given variation ΔQ of the observed quantity Q will be detected only if the related variation ΔI of the image intensity is larger than the total noise level N

$$\Delta I(\mathbf{x_0}) = S_Q(\mathbf{x_0}) \Delta Q(\mathbf{x_s}) \geq N . \qquad (2.8)$$

The total noise can be generated from many factors, including measurement thermal or photon noise, discretization of the object, and truncation error propagation in numerical processing, and strongly depends on the imaging modality. The previous detection criterion is useful in differential imaging for which image variations induced by Q variations are intended to be magnified. Figure 2.6 illustrates the combined influence of the noise and of the limited spatial resolution in such differential imaging.

The dependence of the contrast on the tissue properties explains the complementarity between different imaging modalities. But it is worth noting that the basic sensitivity of the interaction has to be somewhat weighted by the experimental protocols and the subsequent numerical processing in order to define the net global sensitivity of the imaging modality under consideration.

2.2.6 Temperature Dependence of Images

In the hyperthermia context, the parameter temperature T is of prime importance in terms of treatment efficiency. However, noninvasive thermometry is very difficult for at least two reasons. First, the thermal sensitivity of images must be large enough to detect small temperature changes. Second, and more complicated, is the calibration and then the separation between direct and indirect effects of T on the displayed quantity Q. The difficulty of the separation results from the living nature of the observed tissues which are modified in their content by the temperature change resulting from the hyperthermia treatment. A passive body could be characterized once and for all, under certain general conditions, by some temperature coefficient or the variation law of Q with respect to T. For living tissues, some factors other than T, such as the blood flow rate, are implicitly dependent on T, in such a way that the global dependence of Q versus T can be very complicated and subject to extreme variability from one case to another.

It results that the interpretation of images in terms of the temperature only requires substantial knowledge of tissue properties and behavior under temperature changes. Furthermore, since these properties and behavior depend on the nature of the considered tissues, a priori knowledge of the different tissue distribution within the observed volume is necessary. Such knowledge can be achieved by means of already existing tomographic modalities such as X-ray tomodensitometry or NMR imaging.

These first observations show the complexity of the use of imaging modalities for noninvasive thermometry. The factor relating the image intensity and the temperature varies both in space – according to the tissue composition – and in time – according to its evolution. In particular, the expected accuracy of a fraction of a degree seems quite impossible to achieve because of the extreme difficulty of temperature calibration. Probably easier would be to evaluate temperature changes, an absolute temperature scale being

provided by conventional superficial or invasive measurements. But, in any case, the total lack of information on what happens during deep hyperthermia treatments within large heated areas makes imaging modalities the only possible approach to this difficult problem. The information derived from such images is expected to be very useful, both in providing better basic knowledge of phenomena generated by hyperthermia and in making possible better control of hyperthermia sessions.

Two attitudes exist in such complicated situations. The first consists in rejecting a priori imaging modalities because of their expected difficulties and their inability to provide calibrated temperatures. The second, more pragmatically, is oriented toward producing temperature-dependent images and toward their later interpretation by means of suitable training and improvement of the basic knowledge of interaction mechanisms.

2.2.7 Other Practical Considerations

The intrinsic technical performances of a given imaging modality have to be completed by some other important considerations of a practical nature concerning its use in clinical situations. As explained on p. 38, although the spatial and thermal resolutions are of prime importance, they are not sufficient to guarantee practical efficiency. The conditions under which they could be obtained must be specified. The range of measurable temperature variations and the volume over which the temperature could be achieved must be determined. Moreover, the clinical context imposes other constraints such as innocuity, duration of image formation, compatibility with heating equipments, and cost. Innocuity mainly depends on the ionizing properties of the radiation concerned and cannot be underestimated in importance for evident reasons. The necessary limitation of the incident radiation dose is an obstacle to the improvement of the signal to noise ratio during continuous monitoring.

Also quite critical is the time required to get the thermal information, i.e., the time between the instant when measurement is started and the instant when the image is produced. This total duration must accommodate the natural thermal time constants of the observed media. Furthermore, too long measurement times may produce artifacts resulting from more or less unavoidable movements of the patient.

The compatibility with heating equipment is here understood in a very broad sense, including the now classical electromagnetic aspect − the imaging system must not be disturbed by externally or locally parasitic radiation − and more simply the geometrical and mechanical aspects − the shape and the dimensions of the imaging system, as well as its mechanical movements, must accommodate the available space around the patient.

Finally, the economic aspect must be considered, i.e., the cost of development of the imaging modality in question. While the previous constraints can be relatively easily estimated on the basis of objective figures, the cost is more difficult to evaluate. At least two aspects should be taken into account. The first concerns the experimental and preclinical level, for which the price to pay could be quite high but nevertheless acceptable due to the importance of the problem and the ignorance of fundamental mechanisms the knowledge of which is absolutely necessary for the development of deep hyperthermia. The second concerns the systematic integration of the noninvasive thermal control modality in most heating configurations; in this case, it would be reasonable to look at a relative balance between the costs of control and heating equipment. While on the question of costs, it should be noted that fees for hyperthermia treatments are not currently reimbursed in some countries, but demonstration of improved efficacy of hyperthermia, requiring noninvasive control techniques, would probably lead to greater acceptance of this treatment modality.

2.2.8 Optimization of Performances

In an imaging system, the image quality is a function of contradictory tendencies. As already stated, thermal and spatial resolutions and rapidity are not independent. For instance, spatial resolution Δx can be improved by a convenient increase of the measurement duration τ. This is quite evident for diffraction-limited systems, for which the resolution is directly proportional to the wavelength. Since generally the propagation losses are also an increasing function of the frequency, any improvement in spatial resolution is obtained at the cost of a loss of sensitivity which can only be compensated by means of a reduction in the measurement rapidity. For other systems, improvement in the spatial resolution necessitates at least an increase in the number of pixels used to describe the object, implying longer measurement and calculation times. However, longer durations could be compensated by increasing the equipment complexity and cost.

On the other hand, thermal resolution ΔT, or contrast, can be enhanced either by time or space integration. In the first case, the duration is increased. In the second, the integration can be achieved over different reconstructed slices of the object or, in the same plane, over the neighbors of the considered point. Such a space integration causes an evident loss of spatial resolution. The integration can sometimes be done in the Fourier domain, resulting in the same effect.

As a matter of fact, it is the trio Δx, ΔT, and τ which must be optimized. Any improvement of one of these parameters, while keeping the others constant, involves higher equipment cost. Optimization will necessarily include all the specificities of the problem under consideration. It excludes a priori general purpose equipment at the benefit of specific equipment matching in the best possible way a restricted class of constraints.

2.3 Electromagnetic Radiometric Techniques

2.3.1 Presentation

Electromagnetic radiometry measures the natural electromagnetic emissions from the body due to its temperature. The electromagnetic emission spectrum of a black body at human body temperature (37 °C) is shown in Fig. 2.7. This curve can be described mathematically by the Planck expression

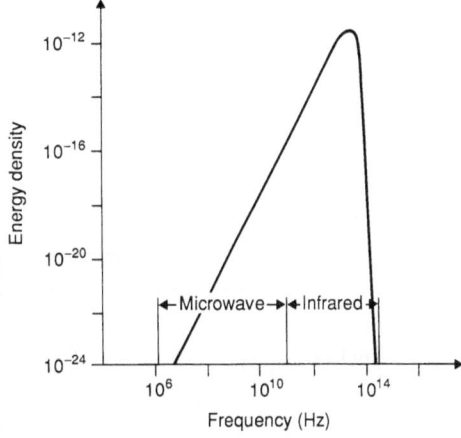

Fig. 2.7. Electromagnetic emission spectrum of a black body at 37 °C (310 °K)

$$P = \frac{2hf^3}{c^2} \frac{1}{(e^{(hf/kT)} - 1)} \qquad (2.9)$$

where: h = Planck's constant, f = frequency, k = Boltzmann's constant, T = temperature, and c = speed of light.

The brightest emission takes place in the infrared portion of the spectrum ($\sim 10\,\mu m$); at microwave frequencies, the power is lower by a factor of around ten to the eighth power. In the microwave region where $f \ll kT/h$, the Planck expression for power emitted from a black body can be approximated by the Rayleigh-Jeans expression

$$P = \frac{2kTf^2}{c^2} ; \qquad (2.10)$$

that is, in the microwave region, the power emitted by a black body is proportional to its physical temperature T. This fact is important for two reasons: First, a body as a whole will radiate into space with a power proportional to its temperature. Consequently, body temperatures can be measured by external power detectors. Second, each elemental volume composing the body, if in thermodynamic equilibrium with its surroundings, will radiate a power proportional to its local temperature. Hence, interpretation of external measurements in terms of internal temperature distributions is possible, a fact which is becoming more important as microwave radiometry techniques become more advanced.

Infrared detection was the first of the two techniques to be used as a clinical tool, in a diagnostic capacity, due to the higher power levels available and more readily available technology. Infrared detection, however, has many drawbacks, perhaps the most serious being the small depth of penetration of infrared radiation in tissue. As a result, infrared measurements only give information on the temperature in a surface layer of tissue up to 0.1 mm thick. Information on deeper lying temperature distributions is only available by means of their referral to the skin surface by conduction or transport mechanisms involving blood flow. Thus infrared methods are naturally limited to controlling very superficial hyperthermia treatments [15]. In contrast, microwaves have much longer wavelengths, from 1 mm to 1 m, leading to much greater penetration in tissue. Temperature is thus available from greater depth. Microwave radiometry has been shown to give information on temperature distributions down to 2 or 3 cm.

Microwave radiometry was developed in the 1940s as a technique for sensing thermal emissions, with applications in radioastronomy and remote sensing of the earth and its atmosphere. Its application to medicine was first proposed by Barrett and Myers [9] in the early 1970s, who employed a contacting waveguide antenna to couple microwave radiation emitted by tissue into a comparison or "Dicke" radiometer working at a center frequency of 3.3 GHz. Figure 2.8 shows a schematic diagram of a typical radiometer design. In the same period Edrich and Hardee [40] developed a millimeter wavelength scanning system employing a remote dish antenna to give images of superficial thermal distributions. Since 1974, developments of these basic radiometer designs have improved the capability of the technique and widened its applicability. Originally, the primary application envisaged was in screening for breast cancer. The technique has since been applied experimentally to a wide range of pathologies with associated raised temperatures, such as inflammation of the joints or rejection of kidney transplants. More recently the technique has been investigated for monitoring of hyperthermia treatments [26]. This section examines

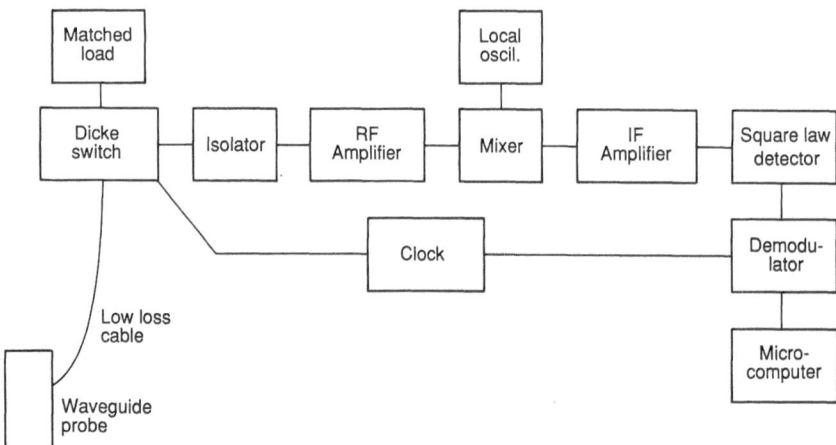

Fig. 2.8. Schematic diagram or a comparison of "Dicke" radiometer

the natural limitations of microwave radiometry and reviews some of the recent advances in this field.

2.3.2 Natural Limitations

The natural limitations of microwave radiometry can be examined in terms of the following parameters:

1. Temperature resolution
2. Measurement duration
3. Spatial resolution
4. Depth of view
5. Interpretation of the measurements in terms of temperature
6. Finanical cost

2.3.2.1 Temperature Resolution and Measurement Duration

The temperature resolution of a radiometer is given by

$$\Delta T = \frac{K(Ts+T)}{\sqrt{(Bt)}} . \tag{2.11}$$

For a Dicke radiometer, $K = 2$. Thus, the temperature resolution is dependent upon Ts, the system noise temperature, T, the temperature of the body, B, the bandwidth of the radiometer receiver, and t, the integration time constant of the final detection stage. The system noise and the bandwidth of a radiometer are a function of system design, and the cost of the radiometer depends very largely on these two factors. System noise is a measure of the degradation of signal to noise ratio between the system input and output and is important mainly in the front-end components, which add noise due to their own physical temperature. The noise temperature of a component is given by

$$Tn = (Fn-1)T_c \tag{2.12}$$

where Fn is the noise figure and T_c the component temperature. Improved noise performance means increased cost of manufacture, although at present costs are falling rapidly due to mass production for the communications industry.

From Eq. 2.11 improvement in temperature resolution can be achieved by increasing the integration time, and a temperature resolution of 0.1 °C is achievable for most present systems with a time constant of 1 s. Any major improvement in temperature resolution will perhaps require cryogenic cooling of front-end components, a technique already widely used in radio astronomy, but one which dramatically increases the cost of a system.

2.3.2.2 Spatial Resolution

Spatial resolution is a function both of the frequency of operation and of the type and size of antenna employed. The resolution of a noncontacting antenna is limited by the wavelength in air. In order to achieve a resolution of around λ on the skin surface, the antenna is normally a few tens of wavelengths in diameter. For a contacting antenna, the resolution is limited by the wavelength in the medium with which the antenna is in contact (λ_m). If the antenna dimension is large compared with λ_m, the spatial resolution is of the order of the size of the antenna. The antenna size can be reduced by propagating the electromagnetic wave in a higher dielectric constant medium within the antenna. For example, an open-ended waveguide antenna can be filled with a very low loss dielectric of permittivity greater than 1. Theoretically, the minimum spatial resolution of an antenna is of the order of $\lambda_m/2$. In practice, the spatial resolution of probes with dimensions around λ_m varies in a complex manner with aperture size. The variation of resolution with probe size can be estimated from the radiation pattern of the antenna into tissue since the antenna is reciprocal in radiation and reception [92]. As an illustration of this dependence, an experimental comparison [61] has been carried out between the spatial resolution of a waveguide antenna when scanning over water (to simulate muscle) and resin (to simulate fat [54]). The antenna has a dimension of 2.5 λ_m in water and 0.6 λ_m in resin. Figure 2.9 shows that the spatial resolution is better when scanned over the high permittivity medium, due to the greater spreading of the probe radiation pattern in the lower permittivity medium.

2.3.2.3 Depth of View

One of the major limitations on microwave radiometry is the poor penetration depth of radiation at microwave frequencies into tissue. Although the penetration depth is many times greater than at infrared frequencies, this factor still severely limits the range of applications for microwave radiometry in monitoring hyperthermia.

The depth of view depends on the penetration depth and on the temperature resolution of the radiometer. The penetration depth depends on:

1. Frequency of operation
2. Tissue type and distribution
3. Antenna type and size

Penetration depth is usually defined as the depth to which a signal can travel in tissue before it is attenuated by a factor $1/e$. Figure 2.10 shows the plane wave penetration depth as a function of frequency for two different tissue types. A wide range of frequencies have been used for medical microwave radiometers. Edrich et al. [39] employed frequencies from 10 to 60 GHz with a noncontacting dish antenna. In this case such high frequencies were necessary to allow an appropriate size of dish and adequate spatial resolution with a noncontacting system, but resulted in low penetration depth. At the other end of the scale, a prototype system recently constructed [19] operates in the frequency band 100–500 MHz. The majority of radiometer receivers operate with a center frequency in the range 1–5 GHz, where the spatial resolution is acceptable for most medical applications. Plane wave penetration depth in this region is from about 1 to 4 cm in muscle, and from 6 to 20 cm in fat. However, such figures can be misleading. For contacting antennas, the size of the antenna has a large effect on the penetration depth (Fig. 2.11). The approximation of plane wave attenuation cannot be employed unless the antenna is of the order of several wavelengths in the medium in its smallest dimension. Various authors have studied the effect of antenna size on penetration depth both in radiometry applications [61] and for heating patterns in hyperthermia [25]. The frequency of operation can be varied over a large range, and if the size of the antenna is kept constant, the penetration depth varies little with frequency. Hence careful antenna design is essential for an efficient system.

Penetration depth varies with the type of tissue over which the antenna is placed. Figure 2.10 shows that penetration depths are much lower in tissues with high water content (e.g., muscle, skin) than in those with low water content (e.g., fat, bone). If more than one tissue type is present, reflection at interfaces can reduce considerably the penetration depth.

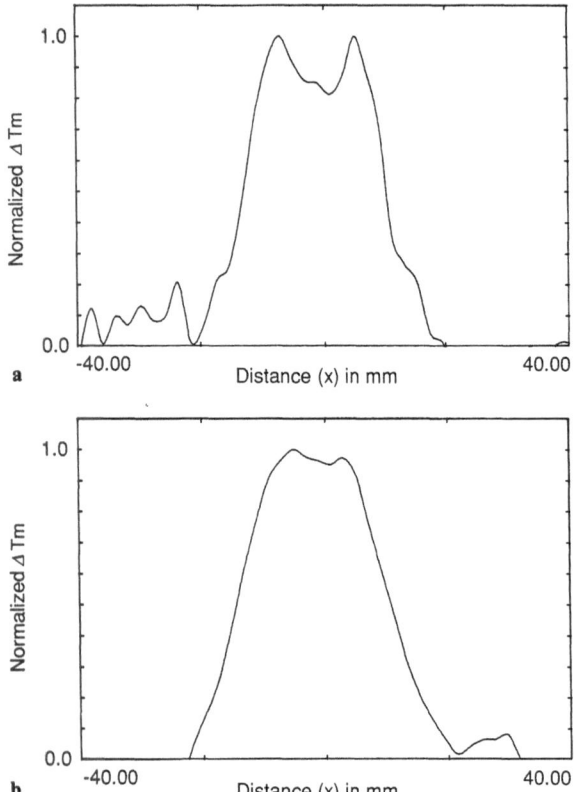

Fig. 2.9a, b. Radiometer output for scanning of a waveguide probe (18 mm×9 mm) over two cylindrical hot spots of radius 9 mm, separated by 3 mm, immersed in water. Frequency: 4.6 GHz. **a** Homogeneous water medium. **b** Waveguide in contact with resin for 9 mm resin and 1 mm water between probe and hot spots

2.3.2.4 Interpretation of Radiometric Measurements

Interpretation of radiometric measurements in terms of temperature is very important for monitoring hyperthermia treatments, where temperature is the critical factor in treatment efficacy. Temperature measurements by microwave radiometry are possible if the radiometer is first calibrated. The usual calibration

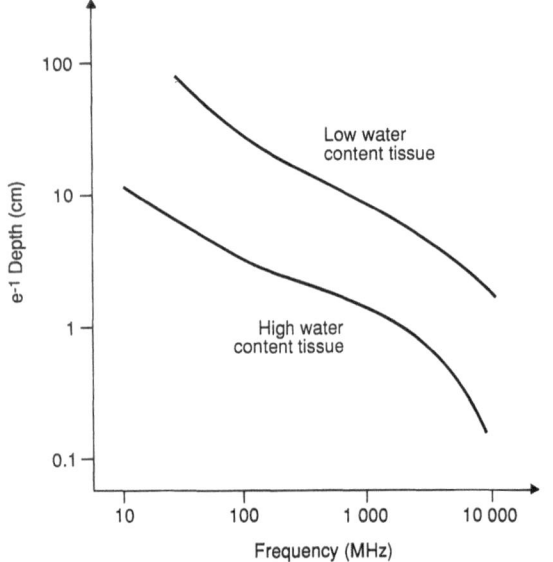

Fig. 2.10. Plane wave penetration depth in tissue versus frequency for two different tissue types

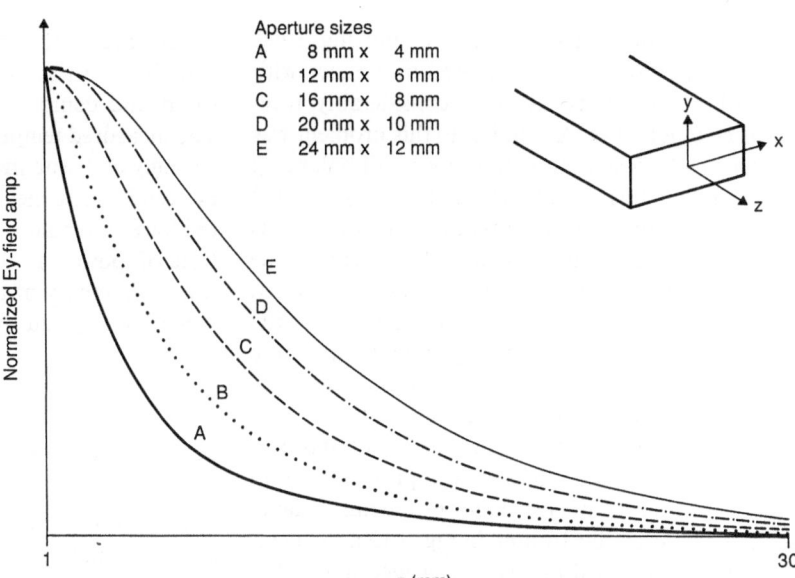

Fig. 2.11. Electric field (y-polarization) on waveguide axis for a range of aperture sizes radiating into water. Frequency = 5 GHz (computer simulated)

method is to measure the radiometer output voltage as a function of water temperature when the probe is in contact with a water bath whose temperature is variable in the range 35°–45 °C and the water in the bath is circulated to keep the temperature uniform. Subsequent measurements by microwave radiometry are generally considered to be direct measurements of temperature and relatively (in comparison to other methods of noninvasive temperature measurement) insensitive to other parameters which may change during the course of a treatment (e.g., blood volume). It is important to examine the extent to which this is true and to assess in which conditions radiometric measurements are appropriate for controlling hyperthermia (see also Sect 2.6.5.8).

Electromagnetic emission from a dissipative body is the result of spontaneous local electric and magnetic moments arising from thermally induced random motions of its constituent particles. To calculate the power radiated into the surrounding medium, the time and space correlation functions of these fluctuating sources must be determined. The fluctuation dissipation theorem [73] describes radiation as being generated by equivalent current sources J(r,ω), with the expectation value

$$\langle \bar{J}(\bar{r}, \omega) * \bar{J}(\bar{r}', \omega') \rangle = \frac{4}{\pi} k \omega \varepsilon''(\bar{r}) T(\bar{r}) \delta(\omega - \omega') \delta(\bar{r} - \bar{r}') \tag{2.13}$$

where ε'' is the imaginary part of the permittivity at position $\bar{r}'$. If we take the emitting region to be a half space and make the approximation that the medium is vertically structured, i.e., that both temperature and permittivity vary with depth (z) only, the brightness temperature emitted by the medium can be written as

$$T_{Bp}(\alpha) = \int_{-L}^{0} W'_p(\alpha, z) \varepsilon''(z) T(z) \, dz \tag{2.14}$$

where α can have the meaning of either frequency or angle of observation, and p selects the polarization. The integral is performed from the surface to a depth L within tissue beyond which the contribution to the brightness temperature becomes negligible.

To examine which factors contribute to the brightness temperature, which is the temperature measured by the radiometer, it is convenient to separate the problem into two cases. Firstly, assume that the temperature of the medium is uniform at temperature T. The brightness temperature of the medium is then given simply by

$$T_B = eT \tag{2.15}$$

where e is the emissivity of the medium either into air or at the interface between the tissue and a contacting antenna, and is a function of permittivity, i.e., of the composition of the tissue. The emissivity of tissue into free space can vary over the body surface from 0.25 to 0.95. As a result, unless areas of the body are chosen for examination which exhibit small emissivity variations (e.g., the breast) measured brightness temperature is primarily a function of tissue emissivity rather than tissue temperature.

Contacting antennas overcome this problem to a large extent. Most antenna types can be matched to a mean value of tissue impedance, say 75 Ω, which diminishes the range of tissue emissivity values to 0.7–1.0. However, important signal variations which are not related to temperature can still result. An additional source of inaccuracy arises if absolute measurements are needed, as in the case of hyperthermia monitoring, since the radiometer is usually calibrated in contact with water whose emissivity can vary considerably from that of certain tissues. A solution to this problem was proposed by Ludeke et al. [77], using a null-balancing radiometer. In this system, an artificially generated signal equal to the measured brightness temperature is injected into the antenna towards the antenna-issue interface. The fraction of this signal which is reflected back into the radiometer exactly compensates for the power lost from the emitted brightness temperature due to the emissivity of the tissues and thus the measurement is a function of temperature only.

When the temperature of the medium is nonuniform, variations in signal due to tissue composition can no longer be entirely compensated for by the null-balancing method. As an illustration, in Fig. 2.12 a temperature hot-spot is buried in a two-layer medium with a fat layer (modeled by Guy recipe resin) overlying muscle (water), and the temperature measured as a function of position as waveguide probe is scanned over

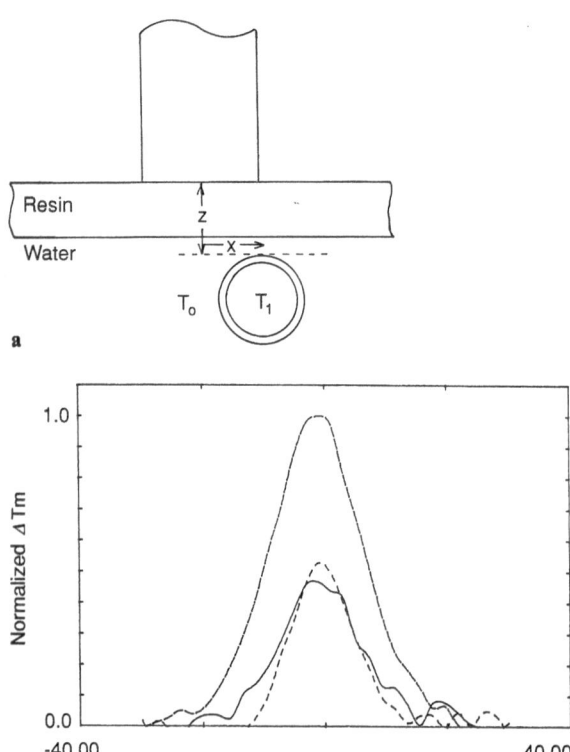

Fig. 2.12 a, b. Radiometer output when scanning a waveguide probe (18 mm × 9 mm) over a cylindrical hot spot (radius 9 mm) immersed in water at a fixed depth ($z = 10$ mm) for three different dielectric structures. Frequency = 4.6 GHz. **a** Setup of waveguide probe and hot-spot. **b** Radiometer output: ———, homogeneous water; – – –, 4.5 mm resin + water; --------, 9 mm resin + water

the surface of the medium is shown. The changes in emissivity of the tissue into the waveguide were measured and compensated for in these measurements. The measured temperature can be seen still to depend strongly on tissue structure. The differences which can be seen in these measurements are due to a balance of the lower attenuation in fat than in muscle, the reflection of power at the muscle-fat interface, and the lower directivity of the antenna looking into fat than when looking into muscle.

2.3.3 Recent Developments

Single probe radiometry, as described in the previous section, is not very successful as an imaging modality for three major reasons. First, the single probe radiometer has no depth resolution and high attenuation in tissue limits the depth of view to fairly superficial tissues. Second, the spatial resolution is not sufficient to allow detailed examination of temperature in small volumes. Lastly, the measurement time (usually greater than 1 s per point) makes scanning of large areas long and difficult.

This section examines some recent developments in the field of microwave radiometry which enhance its imaging capability. These developments can be divided into:

1. Multifrequency systems
2. Multiprobe systems
   a) Noncorrelating systems
   b) Two-probe correlation systems
   c) Aperture synthesis thermography

2.3.3.1 Multifrequency Radiometry

Multifrequency radiometry was introduced in order to improve the depth resolution capabilities of radiometry. As shown by Fig. 2.10, the depth of penetration of radiation into tissue and thus the depth from which information can be gathered decreases as the radiation frequency increases. By taking radiometric temperature measurements at several frequencies, a proportionate amount of information may be collated. There then remains the inverse problem of determining the distribution of temperature with depth from the radiometric data. This problem has been considered by applying various inversion methods; here we shall describe the method proposed by Bardati and Solimini [7]. Beginning from the problem of determining the measured temperature when the temperature distribution is known, Eq. 2.14 can be written as:

$$T_{Bp}(\omega, \hat{k}) = \int_{-L}^{0} W_p(\omega, \hat{k}, z) T(z) dz . \qquad (2.16)$$

The problem becomes that of determining $T(z)$ from a finite and discrete set of measurements of T_{Bp} at various frequencies ω. The technique used by Bardati and Solimini begins by recognizing this equation as a Fredholm integral equation of the first kind. Since the kernel of this equation is ill-conditioned, inversion techniques which exploit the given data in the optimum manner are used, e.g., Kalman filtering. As a first step, the set of weighting functions W_p are computed from a priori knowledge of the tissue structure. An initial thermal profile is computed from a model which takes into account metabolic heat generation, blood flow, and conduction effects. Brightness temperature data corresponding to this profile are computed and compared with measured data. A recursive algorithm then adjusts the thermal profile until computed data and measured data converge. Bardati and Solimini have used simulated data, with added noise to reconstruct temperature profiles (see Fig. 2.13). For applications in hyperthermia monitoring, employing more a priori information on applicator heating patterns and bioheat flow has produced reconstructions which correspond quite well with thermocouple measurements in in vivo animal studies [88].

The ability to retrieve data is limited by physical and cost constraints. The noise associated with the radiometer corrupts data and causes instability in the inversion process and degradation in the spatial resolution. The complexity and cost of a radiometer effectively limit the amount of data which can be measured.

2.3.3.2 Multiprobe Systems

Noncorrelating Systems. A multiprobe radiometer developed in Lille [41] consists of a block of six contacting waveguide antennas in a three by two matrix. Each probe is switched sequentially to a radiometer receiver (Fig. 2.14) operating at 3 GHz. The switching process and data collection are controlled by a microcomputer. Although the present system does not offer any advantage in terms of measurement time, it allows greater precision in the location of the probes than is possible with single antenna systems. This has led to a method for improving the spatial resolution of the imaging process by the overlapping of data sets and suitable definition of probe spatial response. Data processing is performed by a microcomputer and im-

ages are displayed in gray-scale or color on the monitor. Figure 2.15 shows the improvement of spatial resolution achieved in the imaging of a linear hot-spot.

Two-Probe Correlation Systems. Conventional microwave radiometers, as have been considered so far, measure the amplitude of the emitted noise radiation. Correlation systems, introduced more recently into medical thermography [78], extract information concerning the phase content of the signals, thus allowing a much finer spatial location of the signal origin in space and therefore of thermal gradients.

Consider Fig. 2.16 in which two antennas are placed in close promixity in contact with a tissue volume. The antennas collect emitted radiation from volumes which depend on antenna type and size and the tissue characteristics. Volumes V1 and V2 are the volumes coupled to probes 1 and 2 respectively and V3 is the

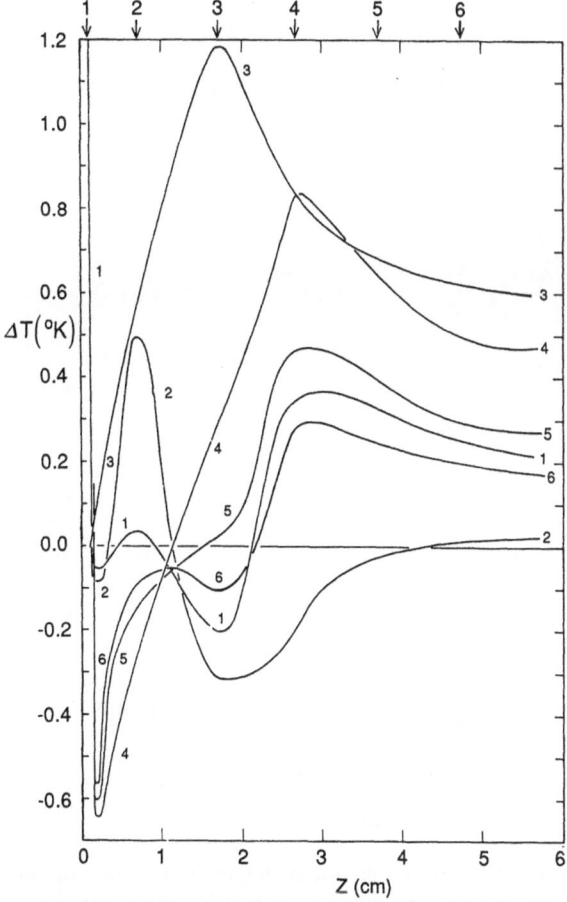

Fig. 2.13. Reconstructed temperature profiles. Average excess temperature retrieved from brightness measurements at 18 frequencies in the range 0.975 – 13 GHz for varying positions of a $\Delta T = 2.2$ K thermal anomaly (position indicated by *arrows at top of diagram*). (By courtesy of F. Bardati)

Fig. 2.14. A 3×2 matrix multiprobe (by courtesy of Y. Leroy)

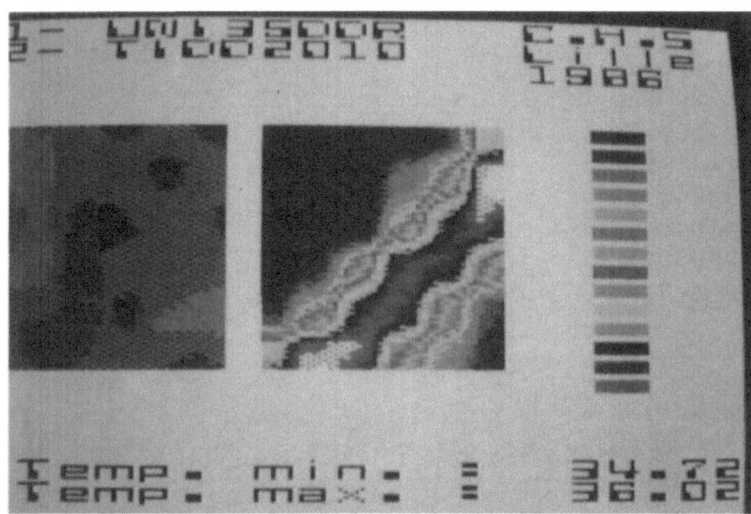

Fig. 2.15. Radiometric image using the multiprobe radiometer of a linear hot spot (by courtesy of Y. Leroy)

volume coupled to both probes. The noise signals originating from volume V3 and reaching radiometers 1 and 2 are correlated, since they originate from the same source, and information concerning the amplitude and relative phase of the signals reaching each probe can be measured. With the switch in position 1, the detected signal originating from a subvolume ΔV_i in region V3 is proportional to

$$C_{i1} + C_{i2} + 2\sqrt{(C_{i1} \cdot C_{i2}) \cos(\varphi_i + \varphi)]\, T_i} \qquad (2.17)$$

and the measured signal is in position 2, the signal from one antenna undergoes a 180° phase change

$$C_{i1} + C_{i2} - 2\sqrt{(C_{i1} \cdot C_{i2}) \cos(\varphi_i + \varphi)]\, T_i} \qquad (2.18)$$

where C_i denotes a coupling parameter between the subvolume and the radiometer probe [equivalent to Bardati's weighting function W(ω, r)] and φ_i is the phase delay between the signal reaching the two probes. φ is a phase delay introduced into one arm of the correlator. The signal is synchronously detected, the output being proportional to the difference in levels, i.e., to $4\sqrt{(C_{i1} \cdot C_{i2}) \cos(\varphi_i + \varphi)\, T_i}$. That is to say that the signal recorded by the correlation radiometer is, in contrast to conventional radiometry, a function not only of the temperature and of the coupling between subvolumes and the probe but also of the relative phase delays between signals reaching each probe. In practice this gives correlation radiometry the advantage of much improved spatial resolution. Figure 2.17 shows the situation when scanning over a buried temperature anomaly.

Still further improvement can be obtained by tilting the guides towards each other. Correlation systems also give the possibility of localizing spatial gradients both longitudinally and with respect to depth, without moving the probes, by means of scanning the phase delay φ in one arm of the correlator [78]. Another interesting feature is that the technique is not sensitive to changes in the uniform temperature of the medium, but only to temperature gradients within the field of view of both probes. Because of this characteristic, correlation radiometry is not viable as a means of monitoring hyperthermia treatment on its own, but could be useful as a complement to conventional radiometry in order to improve mapping of heating distributions.

Aperture Synthesis Thermography. The most recently proposed technique in passive thermography, and one which has not as yet been realized in a practical form is that of aperture synthesis thermography (AST). This technique employs multiple probes to collect thermal data over larger regions than have previously been considered. AST gives improved spatial localization and can lead to much improved spatial resolution. Two different approaches have been considered, either with the utilization of phase information, as in correlation radiometry, or without, as in standard microwave radiometry.

Aperture synthesis thermography with use of phase information was proposed by Haslam et al. [60]. The technique is widely used in radio astronomy, where the fundamental principle is that the distribution of

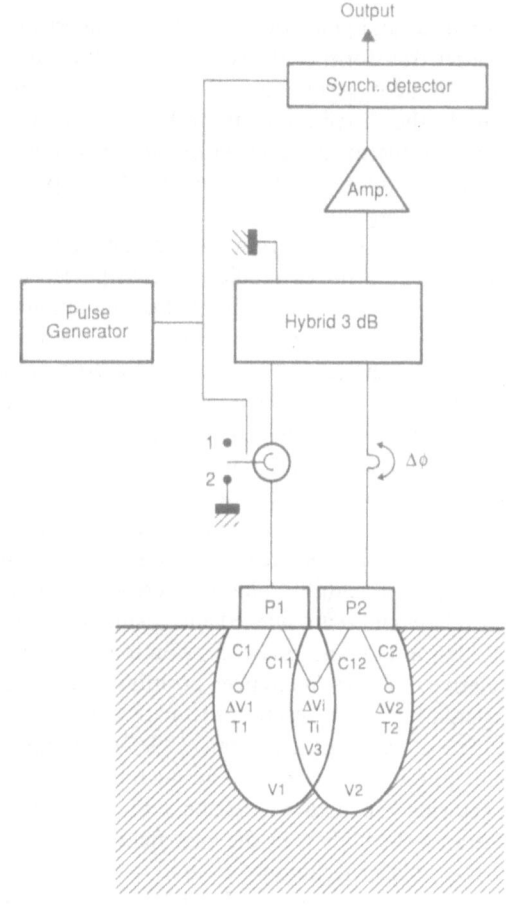

Fig. 2.16. Schematic diagram of a two-probe correlation radiometer. (By courtesy of Y. Leroy)

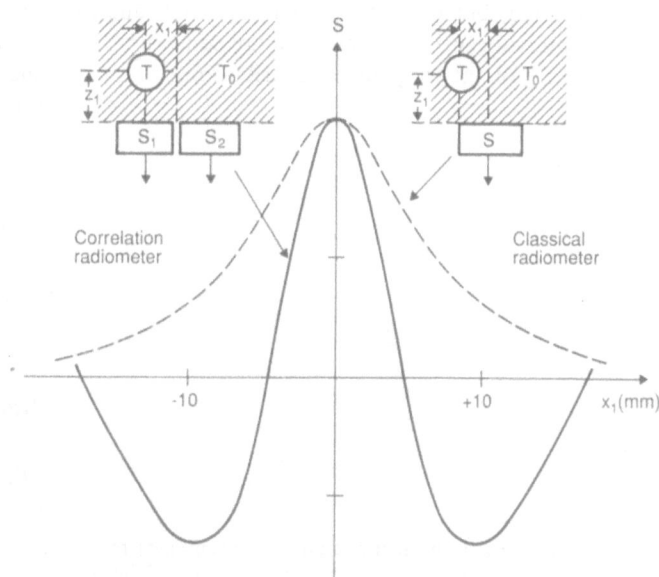

Fig. 2.17. Comparison of radiometric outputs for a standard radiometer and a correlation radiometer for scanning over a cylindrical hot spot immersed in water. (By courtesy of Y. Leroy)

amplitude and phase over an antenna's aperture is the Fourier transform of the source's brightness distribution. In radio astronomy, a line of antennas is used to sample the distribution, the other coordinate being sampled due to the rotation of the earth. Haslam et al. have proposed a system for thermographic mapping over large areas of the human body in which the microwave emission from the human body is measured by an array of dipole antennas placed in close proximity to the body (Fig. 2.18). The antennas are coupled to the body by means of a contoured bed with a high dielectric constant, which serves to reduce problems of signal reflection at interfaces and to improve spatial resolution. The antenna is either a full two-dimensional array or a linear array moveable in a second dimension (linear or circular translation). The receivers are correlation radiometers, measuring the cross-correlation of the signal from each dipole with that from a centrally placed reference antenna. The correlators measure sine (S) and cosine (C) terms, and the amplitude and phase distribution over the array relative to the reference position are given by calculating, for each antenna, amplitude $A = \sqrt{(S^2 + C^2)}$ and phase $\varphi = \tan^{-1} (S/C)$. The amplitude and phase are relayed to the microcomputer, which performs the data processing and gives an image. Because the body is in the near field of the synthetic array, Fourier techniques as used in radio astronomy are, in fact, not appropriate in this case, and use of Fresnel transforms has been proposed. The signals received by each antenna pair will not have equal path lengths in the body and will therefore suffer unequal attenuation. Before a Fresnel transform can be performed, compensation for this effect must be carried out. Prior calibration using phantoms may be a partial solution, but will lead to imaging errors due to tissue inhomogeneity and may require some a priori knowledge of the temperature distribution.

A full analysis of system performance characteristics and requirements, and of the problems to be faced in constructing such a system, is given in Haslam et al. [60]. Only a brief outline will be provided here. The area of the body which can be imaged is limited by the requirement that the whole area must be in the primary beam of all the individual antennas, and is expected to be around 50 cm in diameter, with a theoretical spatial resolution of about 5 mm at 3 GHz and 1.5 cm at 1 GHz. The choice of operating frequency, as in the case of conventional radiometry, is a compromise between spatial resolution and depth of view. To achieve the desired depth of view for this system, a temperature resolution of a few millikelvin is required, which is a factor of 100 times better noise performance than most medical microwave radiometers currently in use, although such high-performance radiometers exist in radio astronomy. It is expected that the system will be able to measure a hot-spot of 1.5 °C above its surroundings at a depth of 5.2 cm in tissue with measurement times of between 6 s at 915 MHz and 222 s at 3 GHz.

A synthetic array which does not take advantage of correlation has been proposed by Bardati et al. [8]. With reference to Eq. 2.16, an array of multifrequency radiometers measures brightness temperature as a function both of the frequency ω and of the direction of propagation $\hat{k}$, increasing the amount of data available for the inversion process over that available from a single multifrequency radiometer (see Sect. 2.3.3.1). The inversion is performed in much the same way as for the single radiometer, giving information on the transverse temperature distribution as well as the distribution with depth. Figure 2.19 shows a simulated reconstruction of a two-dimensional (one lateral dimension and depth) temperature distribution through inversion of radiometric data at ten microwave frequencies for an array of overall length 1 m placed at 50 cm above the skin surface. Spatial resolution is about 1 cm in depth but is much worse in the lateral dimension. These simulations assume measurement noise at a level of 10^{-4} K. Increasing the noise towards more realistic levels results in growing instability of the reconstructions, and poorer spatial resolution.

2.3.4 Discussion

Microwave radiometry is the most advanced of all the techniques considered in this chapter in terms of the length of time for which it has been used for noninvasive temperature measurement. It remains the only

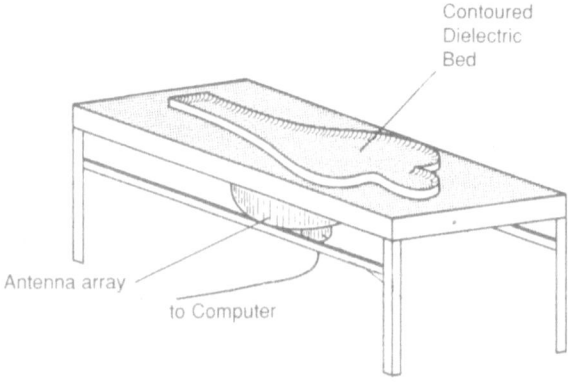

Fig. 2.18. Schematic diagram of a possible arrangement for an aperture synthesis thermogram (by courtesy of A.E. Gillespie)

technique to have been integrated into a commercially available hyperthermia system [27]. As a result of the amount of research which has been carried out in this field, many of the advantages and drawbacks to the use of radiometry for hyperthermia monitoring are now well recognized.

One of radiometry's greatest advantages is that it is passive, with no possible damage resulting to the patient from its use. In addition, certain configurations can be found for which radiometry is compatible with heating equipment. For example, if the same frequency is used for microwave heating and detecting, the same antenna can also be used and switched between heating and receiving.

Among the disadvantages of employing radiometry, one of the most serious is, perhaps, the limited depth of view. Radiometry will receive temperature information from depths of no more than a few centimeters even at a low center frequency. High attenuation in tissue also means that, although the temperature resolution is 0.1 °C at the tissue surface, the temperature resolution at 1 cm depth can be of the order of a few degrees.

A measurement duration of several seconds is acceptable if only one data point is needed. However, when collecting a number of points, data acquisition times become long, possibly resulting in inefficient heating sequences since the heating must be switched off during data collection.

The spatial resolution is adequate for a hyperthermia monitoring technique. The lateral spatial resolution is predominately a function of probe size, and the small depth of view effectively places an acceptable limit on the depth resolution. Recent advances in microwave radiometry serve to improve spatial resolution. Multifrequency devices improve depth resolution, and correlation radiometry can improve resolution in both dimensions, especially the lateral plane, where the resolution can be as good as 1 mm.

A microwave radiometer with a single probe working at a single frequency records only one data point at a time, as does a single thermocouple. An advantage of the thermocouple is that its position is more or less well known and this position can be chosen to give the optimum control conditions. In the case of a radiometer, an average temperature is collected, the average being over a relatively large volume (say $2 \times 2 \times 2$ cm at 3 GHz) and weighted by a rapidly decaying exponential function, which depends on the tissue types present, in favor of the temperature at very superficial levels. Such data can be useful, but care must be taken in interpreting the measured temperature when the efficiency of a treatment is thus controlled.

Interpretation of radiometric data is a difficult task since both the emission characteristics and the attenuation of the signal depend on tissue types present within the field of view of the probe. A priori information on tissue composition would facilitate such interpretation. Additional problems result from the dependence of the measured temperature on parameters other than tissue temperature, specifically on the dielectric complex permittivity $\varepsilon' - j\varepsilon''$. Heating of tissue by hyperthermia can change its permittivity for the duration of the treatment or, indeed, permanently. Variations in blood supply to the heated tissue, as a result of the body's thermoregulatory system, can greatly influence both ε' and ε'', and therefore create the false impression of a change in tissue temperature. It is not clear how such changes could be taken into account.

In conclusion, conventional radiometry has been shown to be useful as a general indicator of volumetric temperature changes confined to fairly superficial tissue regions. Advantage can be gained by operating a single antenna for heating and signal reception to allow the signal to be collected from the same region as is being heated. As a note of caution, if it is necessary to employ a bolus in the above situation to prevent skin burning, it should be realized that the radiometer will primarily measure the temperature of the bolus and not the tissue. Clinical experience has also shown that, since the radiometer is a broad-band noise detector, the device is very sensitive to environmental electromagnetic radiation, for example airport

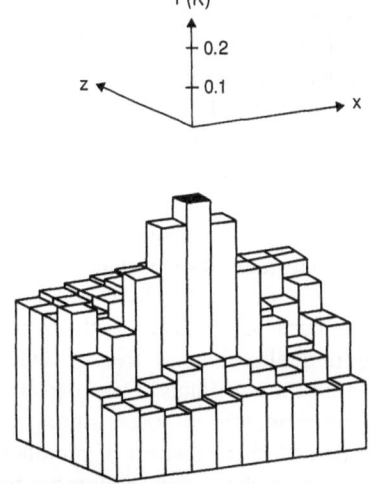

Fig. 2.19. Reconstruction of a 1 K excess temperature in a multilayer tissue medium through inversion of radiometric data at ten microwave frequencies in the band 0.975 – 11.5 GHz. The location of the hot pixel is denoted by the *black square*. (By courtesy of F. Bardati)

radar. It is often necessary to operate the radiometer inside a Faraday cage.

Recent developments in microwave radiometry have improved the imaging capability of the technique. Multifrequency radiometry allows depth information to be collected, giving, for the first time, some depth resolution. Correlation radiometry considerably improves spatial resolution, although since the received signal is sensitive solely to temperature gradients, this technique can only be used in conjunction with other temperature measurement techniques for monitoring hyperthermia. Aperture synthesis thermography gives microwave radiometry a true imaging capability. The envisaged system is expected to give reasonable temperature resolution and depth of view for data acquisition times which are acceptable. The design constraints imposed to achieve these figures are, however, very strict, requiring state of the art, and thus very expensive, technology. It remains to be seen whether the technical and theoretical problems can be overcome and, if so, whether such a system can be constructed at acceptable cost.

2.4 X-ray Tomodensitometry

2.4.1 Presentation

X-ray tomodensitometry is probably the most popular tomographic technique in the biomedical area. The success of its development has resulted from two favorable conditions. First, X-ray tomodensitometry has been supported by many decades of classical radiography practice, and hence has profited from invaluable clinical experience. Secondly, due to their specific mode of interaction with living tissues, X-rays benefit from two "linear" properties: (1) they propagate according to linear trajectories, and (2) tomodensitometry can be related to a linear (previously called quantitative) imaging problem with respect to tissue densities. These two favorable conditions have been responsible for the rapid procurement of good quality images free of shadowing effects or propagation artifacts. They also explain why this imaging modality has been among the first to be investigated in the context of remote thermometry. Existing and elaborated equipment has made possible the evaluation of its potential in realistic situations. After a brief survey of the principles of tomodensitometry, the main causes of performance limitations are analyzed. Next, the particular application to

remote thermometry is addressed and some results are presented.

2.4.2 Fundamentals

As explained by its name, X-ray tomodensitometry provides cross-section views of tissue densities. The main feature making this possible is that X-rays are absorbed by tissues according to their density. The local linear absorption coefficient at point P, $\mu(P)$, is proportional to the density ϱ:

$$\mu(P) = K \cdot \varrho(P) \ . \tag{2.19}$$

Saying that μ depends on P implicitly means that the density depends on the tissue nature, composition, temperature, etc.

The CT number has been introduced as follows:

$$CT = 1000 \ (\mu/\mu_w - 1) \tag{2.20}$$

where μ and μ_w represent, respectively, the absorption coefficient of the tissue and the absorption coefficient of water at room temperature. CT numbers are expressed in Hounsfield units (HU). As shown later, μ and, consequently, CT numbers are a function of the effective energy of the X-ray source. This dependence is especially significant for tissues with large atomic numbers Z, such as adipose tissues, for which photoelectric absorption is high. The sign of variations of CT numbers with energy depends on the tissue composition.

CT numbers can be calculated theoretically. Table 2.2 gives some examples of CT numbers corresponding to different soft and hard tissues, illustrating the contrast capabilities of X-ray scanners with respect to tissues of similar or different kinds.

With X-rays, collimated beams can be easily produced in such a way that the absorption along a ray trajectory is simply related to the density of the tissues lying on this trajectory. In the experiment depicted in Fig. 2.20, where the beam is assumed to be located in the

Table 2.2. CT numbers of different tissues [79]

Tissue	Calculated CT number
Adipose	-92
Heart	22
Brain	32
Kidney	42
Pancreas	44
Blood	54
Spleen	55

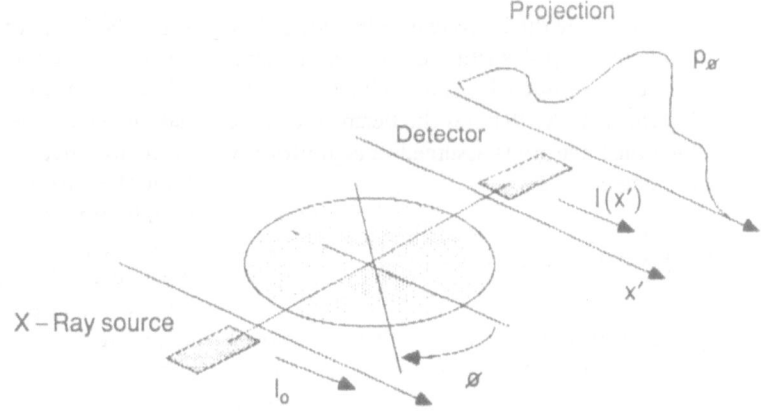

Fig. 2.20. Basic configuration for X-ray tomography

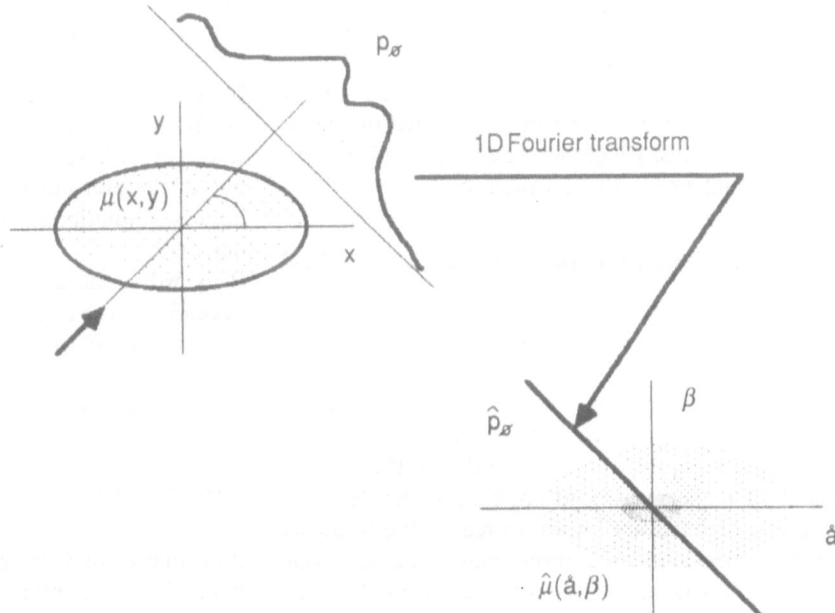

Fig. 2.21. Projections and associated Fourier domain

plane $z = 0$, the ray intensity $I(x')$ is given as a function of the emitted intensity I_0 by the following integral:

$$I(x') = I_0 \exp\left\{ -\int_L \mu(x, y)\, dl \right\} \qquad (2.21)$$

or, equivalently, introducing the projection p_θ for the aspect angle:

$$p_\theta = -\text{Log}\,\{I(x')/I_0\} = \int_L \mu(x, y)\, dl \ . \qquad (2.22)$$

Taking the Fourier transform $\hat{p}_\theta$ of a set of projections p_θ allows to fill in the Fourier plane of μ (Fig. 2.21).

In practice, a fan beam is mechanically rotated with an array of fixed or rotable array of detectors. A complete rotation takes approximately 2–10 s, depending on the generation of the scanner (translation/rotation, rotation only, etc.). The number of recorded views is of the order of 1000–2000. Data processing can be achieved by using different techniques which will not be detailed here [90]. The pixel size is ca. 1 mm.

2.4.3 Image Quality

As already mentioned, the linear absorption coefficient μ is ideally only dependent on the coordinates x and y. As a matter of fact, the dependence of μ versus z and the energy E of the incident beam should be compensated for. More explicitly, the finite thickness and the polychromaticity of the incident beam must be considered as factors limiting the image quality. These dependencies introduce nonlinearities of the measurement with respect to the density (e.g., [11, 45]).

The effect of the finite thickness can be introduced by means of a more careful writing of the relationship between the recorded intensity and the linear absorption coefficient μ. If Δz denotes the beam thickness, within which the intensity is assumed to be uniformly distributed, one can write:

$$I(x') = I_0 \left| \int_{\Delta z} \left\{ \exp \left\{ -\int_L \mu(x,y)\,dl \right\} \right\} dz \right| \qquad (2.23)$$

The projection is then no longer directly proportional to μ, and the nonlinearity is more especially important as the tissue composition varies in the beam cross-section. It is not possible to proceed to some correction and the only possibility is to produce beams as thin as possible.

The situation is slightly different for the energy dependence of μ. If the beam is described by means of a photon density N(E) in the energy band ΔE, the measured intensity can be written as:

$$I(x') = \int_{\Delta E} N(E) \exp \left\{ -\int_L \mu(x,y,E)\,dl \right\} dE . \qquad (2.24)$$

Usually, the X-ray sources for biomedical applications produce a spectrum between approximately 30 and 120 keV. The absorption coefficient μ is a decreasing function of E. The mean energy of the transmitted beam is consequently higher than that of the incident beam. The beam is said to become harder. The resulting artifacts are known to reduce the influence of the central area under investigation as against that of the peripheral area. Even with a uniform phantom, the image is deformed, the central part being less intense than the boundaries. The nonlinearity of the μ dependence makes rigorous compensation difficult. In the best case, some calibration can be introduced from projections obtained with a phantom of known properties.

In addition to these two nonlinearities, the image quality is limited by noise considerations. For usual equipment, the preeminent noise is a photon noise. In classical imaging systems, the study is relatively simple due to the fact that the noise at each point of the image is equal to the square root of the number of the received photons. In tomographic imaging, the noise at a given point depends, in complex fashion, on all the points of the object, in such a way that the variance is increased and the signal to noise ratio is reduced. Nevertheless, the signal to noise ratio can be explicitly calculated at the center of a uniform disk of radius a, with absorption μ, as follows [44]:

$$S/N = \mu\, a \exp(-\mu a/2) \sqrt{N_0 n^{-3/4}} \qquad (2.25)$$

where N_0 is the total number of received photons and n is the number of pixels in the disk area. This simple relationship allows some important remarks on the parameters. First, considering that the smallest detectable relative absorption $d\mu/\mu$ is nothing other than the inverse of the S/N ratio, it appears that, maintaining the same spatial resolution n, $d\mu/\mu$ varies as $\sqrt{N_0}$, and thus varies slowly with the incident dose. This incident dose is, on the other hand, limited for safety reasons. Second, obtaining a better spatial resolution n at a constant signal to noise ratio implies that the dose is increased as n^3. Nevertheless, such an increase in the dose must be necessarily limited. Possible solutions to such a dose limitation can be achieved by nonuniform distribution of the dose for all the projections, or by nonuniform distribution in the same projections. These solutions could be efficiently used if only a small area of interest has to be covered. The dose is then concentrated by using one of the previous solutions in this area, preventing useless irradiation in the remaining part of the scanner field.

It is worth noting that the signal to noise ratio depends on the incident energy E. Indeed, as the absorption coefficient μ is itself energy dependent, it is not too complicated to show that the signal to noise ratio is maximum at the energy E_M such that:

$$\mu(E_M) = 2/a . \qquad (2.26)$$

As a matter of fact, this is of little consequence for living tissues exhibiting almost identical mean absorption coefficients or for large targets ($a \# 20$ cm) for which the Compton effect is the prime cause of absorption. The maximum of the signal to noise ratio is centered around 240 keV but is very "flat" and hence not very critical.

In addition to these fundamental reasons of physical orders, the image degradation can be explained by some considerations relevant to incomplete data collection (shadowing) or processing (inhomogeneous filling of elementary pixels). As a result, it appears that, for typical equipment, the short-term estimation of the fluctuation of the CT number of homogeneous and known target fluctuates from 3 to 8, reducing the contrast discrimination in noninvasive thermal sensing. As well as these short-term fluctuations, longer term drifts due to electronic instrumentation must be taken into account, at least for the earliest equipment. Finally, the spatial resolution is approximately given by the elementary pixel size, that is to say, it is of the order of magnitude of 1 mm. The slice thickness can be varied between approximately 1 and 10 mm.

2.4.4 Temperature Dependence of CT Numbers

CT numbers are temperature dependent because of the bulk thermal density of biological materials. It is not without interest that when CT scanners were first introduced such a temperature dependence was considered a possible source of error in high-precision quantitative computed tomography [22]. In 1979, it was demonstrated that the thermal expansion of living tissues was able to induce observable effects [44, 100, 101].

Using the definition equation of the CT number, and introducing the temperature dependence of the density, gives:

$$\frac{d(CT)}{dt} = -\alpha(T)\{CT(T)+1000\} \qquad (2.27)$$

where the thermal expansion coefficient $\alpha(T)$ is given by:

$$\alpha(T) = -\frac{d\{\text{Log }\varrho(T)\}}{dt} . \qquad (2.28)$$

As an example, the thermal expansion coefficient for water at $37\,°C$ is approximately $3.6 \cdot 10^{-4}/°C$, resulting in a CT decrease of the order of $-0.36/°C$. A temperature increase of $1\,°C$ produces a decrease of -0.36 CT number. Such a decrease should be directly compared with the "standard" fluctuations extending from 3 to 8 CT numbers. The comparison immediately shows that conventional scanners cannot be used unless their fluctuations are decreased for a factor greater than 10 in the best cases.

The noise reduction can be achieved by different means. The main idea is to integrate, in the time and/or in the space domains, in order to improve the signal to noise ratio. An immediate consequence is that the measurement duration or the space resolution will be degraded. Spatial integration can be achieved over several slices, or in one given slice. Integrating over several slices leads to the previously described nonlinearities, with the risk to smooth temperature gradients from one slice to another. Alternatively, considering one single slice leads to integration over more than one pixel area, and then to reduced spatial resolution. The general formula giving the signal to noise ratio allows us to predict the contrast improvement resulting from such an integration.

Due to the $n^{-3/4}$ dependence, the standard fluctuation σ_N, after integrating over N pixels, is related to the standard fluctuation σ over one pixel by the following formula [44]:

$$\sigma_N = \sigma\{1+N\}^{-3/4} . \qquad (2.29)$$

Such a formula conveniently describes the observed fluctuations of the CT number with N. Starting from an initial standard deviation $\sigma = \sqrt{40}$ with a pixel area of $2.5\,\text{mm}^2$, a spatial integration over N = 100 pixels will provide a standard deviation $\sigma_N = 30$. The initial temperature resolution of approximately $9\,°C$ over $1\,\text{mm}^2$ is changed for $2\,°C$ over $1\,\text{cm}^2$. This formula clearly illustrates the need for a compromise between spatial and thermal resolutions. The spatial integration can be achieved on the scanner image, after tomographic processing. It could also be done during the processing as filtering in the spatial frequency domain.

2.4.5 Some Results

Some experiments have been conducted on phantoms and on excised tissues. In all cases, an image subtracting technique has been used in order to "magnify" image deformation under temperature changes. The order of magnitude of water CT thermal coefficient $- 0.40$ HU/°C $-$ was confirmed with 99.9% confidence (Fig. 2.22) [44]. Similarly, in vitro experiments on muscle tissue yielded a thermal coefficient of 0.45 HU/°C with 99.8% confidence. A temperature resolution of between $0.25°$ and $0.50\,°C$ was obtained with a spatial resolution of $2.5\,\text{cm}^2$.

More sophisticaled results have been published, including isotherm contour mapping in a phantom [12]. The phantom consisted of a Plexiglas 200 mm diameter cylinder with two different compartments filled with water and 4% agar-agar, respectively. Figure 2.23 illustrates the phantom geometry. A water bath was assumed to create a temperature gradient in

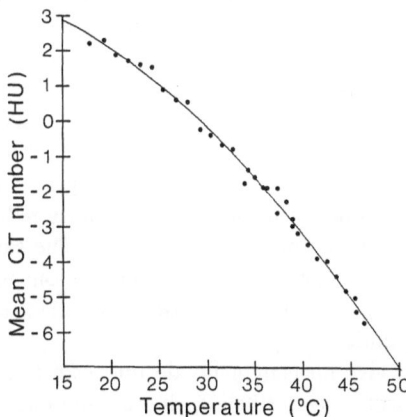

Fig. 2.22. Water CT number variations with temperature. (By courtesy of S. M. Bentzen)

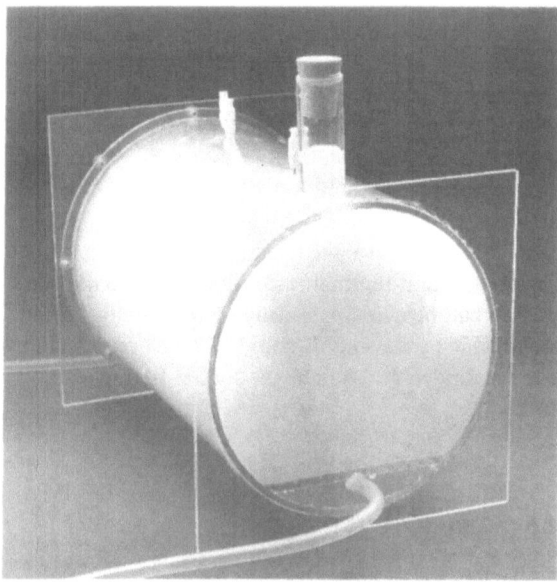

Fig. 2.23. Water/agar-agar phantom geometry for X-ray noninvasive thermometry. (By courtesy of S.M. Bentzen)

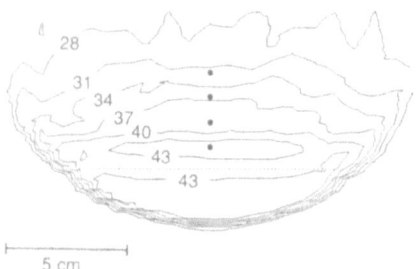

Fig. 2.24. Isotherm lines deduced from X-ray tomodensitometry for the phantom illustrated in Fig. 2.23. (By courtesy of S.M. Bentzen)

the agar-agar compartment without too much heat convection. After some empirical calibration of CT numbers versus temperature and convenient linear interpolation, isotherm contours were displayed and compared with local invasive measurement with thermocouples with rms deviation of less than 0.5 °C (Fig. 2.24). The spatial resolution in the Fourier filtered difference image was ca. 4.2 mm. The time required to produce such isotherm lines is approximately 10 min but could probably be significantly reduced by using higher performance computers. The same technique has been applied to test a four sections balanced coaxial applicator within a 10-cm muscle tissue equivalent cylinder.

In vivo measurements do not seem to have been performed. Nevertheless, the thermal coefficient for living tissues is expected to be at least five times the water coefficient [17].

2.4.6 Discussion

X-ray tomodensitometry appears an interesting approach to noninvasive thermometry. The thermal and spatial resolutions which have been obtained are really very attractive. The main key to the successful results previously recorded is probably the availability of existing and sophisticated equipment which eliminates basic technical problems. On the other hand, it is hard to imagine that the present performances will be drastically improved in the near future and it is clear that some specific or general difficulties limit the use of X-ray tomodensitometry in clinical situations.

As specific difficulties, the following must be mentioned. Firstly, X-ray tomography is an ionizing modality. Basically, this is a serious limitation both to performance improvements − through the signal to noise ratio − and to routine and continuous monitoring of patients, even if one considers possible strategies to distribute the incident dose in the area of interest.

Secondly, the problem of compatibility of the imaging equipment with heating electrodes and applicators has not yet been completely solved. The available space inside a scanner field is rather limited. Furthermore, an alternative approach combining sequential heating and imaging steps, with corresponding movements of the patient, opens the door to repositioning uncertainties as well as to electronic instabilities.

Thirdly, at present no real-time operation seems possible with existing scanners. The fundamental low sensitivity of thermal coefficients with temperature requires improved data processing. But, at midterm, it could be expected that the simultaneous development of efficient algorithms and numerical processor capabilities will be able to overcome this drawback.

In addition there are general difficulties related to the use of any imaging modality for noninvasive thermal purposes. The CT thermal coefficient is evidently different for different tissues, and any calibration attempt in order to achieve absolute thermometry must include a priori knowledge of these coefficients. Although not completely insoluble, this problem is still complicated by the well-known fact that, during hyperthermia, the tissue nature can be altered under the effect of the phenomenon to be investigated. Blood flow rate increases and edema formation are classical examples. Undoubtedly, better knowledge of the real influence of these modifications must be investigated via in vivo experiments.

Nevertheless, prior to realization of the desired improvements in X-ray tomography for clinical use, some immediate applications can be considered when invaluable improvement is achieved as compared with other existing modalities. Using phantom materials with larger thermal coefficients allows the noninvasive thermal analysis of homogeneous media, for instance in order to compare the efficiency of applicators of different kinds. With a CT number of approximately -187 HU and a thermal coefficient of roughly $1.24 \cdot 10^{-3}/°C$, a sensitivity of the order of 1 HU/°C – that is to say, fivefold that of water – can be expected at 20 °C for benzene. Table 2.3 gives some CT values and thermal sensitivities for different materials. For such thermometry in simple phantom configurations, X-ray tomography seems to be a very efficient and accurate tool of investigation.

2.5 NMR Tomography

2.5.1 Fundamentals

The development of nuclear magnetic resonance (NMR) imaging has taken place with remarkable speed over the last few years. Despite the high financial cost of these imaging machines, many hospitals worldwide now have an NMR facility. The image quality of NMR is as good as, if not better than, the longer established X-ray computerized tomography (see preceding section) and has several advantages over this technique. NMR imaging is more flexible, having the ability to measure a number of different parameters and to provide images of any desired plane within the body, oriented in any direction. Also, using nonionizing radiation renders NMR much less hazardous. On the other hand, disadvantages include longer data acquisition times and higher unit cost.

Investigation of NMR for monitoring hyperthermia treatments began in 1985. One of the properties imaged by NMR, the spin-lattice relaxation time constant T_1 of the proton, has been demonstrated to be

a function of the temperature in its immediate environment.

Nuclear magnetic resonance imaging is based on interactions of applied magnetic fields and radiofrequency (RF) radiation with the natural magnetic moments of atoms in the human body, most commonly with that of the nucleus of the hydrogen atom – the proton. The proton has a magnetic moment due to its spin and, when an external magnetic field B_0 is applied to a system of protons (e.g., the human body), the magnetic moments precess around the direction of the magnetic field, some with their axes pointing in the direction of the field (lower energy state), and some with their axes pointing in the opposite direction (higher energy state). In normal conditions there are a greater number of protons in the lower energy state, and a net magnetization M_0 exists. The frequency of rotation of this precession is given by

$$\omega_0 = \frac{\mu B_0}{J} = \gamma B_0 \qquad (2.30)$$

where μ is the magnetic moment of the proton and J is its angular momentum. ω_0 is known as the Larmor frequency (42.6 MHz at 1 T). When RF radiation at frequency ω_0 is applied to this system, more protons absorb energy by transferring to the higher energy state than are stimulated to fall to the lower energy state and thus the net magnetization is changed.

After the RF radiation has been switched off, the system of protons reverts (or relaxes) back to its equilibrium configuration by emitting RF radiation at the Larmor frequency. The relaxation processes depend on the environment of the protons and have associated time constants (of the order of tens of milliseconds) which can be deduced from measurements of the intensity of the emitted radiation with time. The two principal relaxation times are the spin-lattice relaxation time T_1, due to energy exchange between proton spin with the surrounding lattice, and the spin-spin relaxation time T_2, where spins exchange energy amongst themselves. It has been found that T_1 is highly dependent on tissue type and so this quantity is often measured to give high-contrast images.

Several schemes exist for extracting spatial information from NMR measurements. If a linear gradient is imposed in the magnetic field in a given direction, the absorption frequency, from Eq. 2.30, is a function of the field and therefore of position in this direction. If a small RF bandwidth is used for irradiation and detection of the relaxation signal, a "slice" of the object can be selected (see Fig. 2.25). One of the first methods for NMR scanning was to apply magnetic

Table 2.3. CT numbers and thermal sensitivities at 20 °C (Bentzen et al. [12])

Material	CT	a	$\{\Delta(CT)/\Delta T\}/°C$
Benzene	−187	1.237	1.01
Ethanol	−220	1.120	0.87
Glycerol (80%)	171	0.505	0.59
Calcium chloride (40.9%)	988	0.458	0.91
Sodium chloride (20.6%)	262	0.414	0.52
Water	0	0.207	0.21

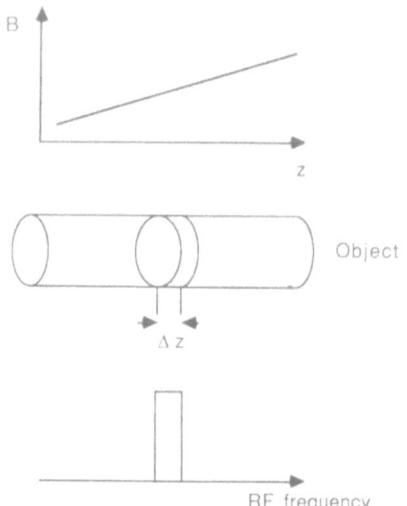

Fig. 2.25. Principle of slice selection in NMR imaging

field gradients in three dimensions to select a given point in the object for examination. However, this procedure required of the order of 1 h to produce a 128×128 single slice image of the body.

A much quicker method of reconstruction is to measure the decay signal as a function of time after a pulse of RF radiation. If a magnetic field gradient G_x is induced in the x direction, Eq. 2.30 becomes

$$\omega_x = \gamma B_0 + \gamma x G_x \qquad (2.31)$$

which shows the relationship between the signal frequency and position in a linear field gradient. For field gradients in three dimensions, the signal decay with time after an RF pulse (known as the free induc-

tion decay or FID) is, for generalized field gradients $G_x(t)$, $G_y(t)$, $G_z(t)$

$$S(t) = K M_0 \iiint_{xyz} \varrho(x, y, z)$$

$$\times \exp \left[i \gamma \int_0^t (x G_x(t') + y G_y(t') + z G_z(t')) \, dt' + i \gamma B_0 \right]$$

$$\times \exp(-t/T_2) \, dx \, dy \, dz \qquad (2.32)$$

where K is a constant depending on receiver coil design and the electronic apparatus and ϱ is the proton density. If we apply the previous scheme to select a slice of the object, the problem is reduced to two-dimensional slice scanning. The Z gradient is applied at the time of the RF excitation to select the slice and gradients G_x and G_y are applied during the relaxation period (Fig. 2.26). Equation 2.32 reduces to

$$S(t) = K M_0 \iint_{xy} \varrho(x, y, z) \exp[i \gamma (x G_x + y G_y) t]$$

$$\times \exp(-t/T_2) \, dx \, dy \qquad (2.33)$$

writing $r = x \cos\theta + y \sin\theta$ where θ is the angle the composite gradient makes with the y-axis,

$$S(t) = K M_0 \int_r I_1(r) \exp(i \gamma r G_r t) \exp(-t/T_2) \, dr \qquad (2.34)$$

where $I_1(r)$ is a line integral to the spin distribution perpendicular to the applied composite gradient G_r and is obtained by a one-dimensional Fourier transform of S(t). The line integral projections can be made to rotate around the object to give a series of views. This situation is analogous to X-ray com-

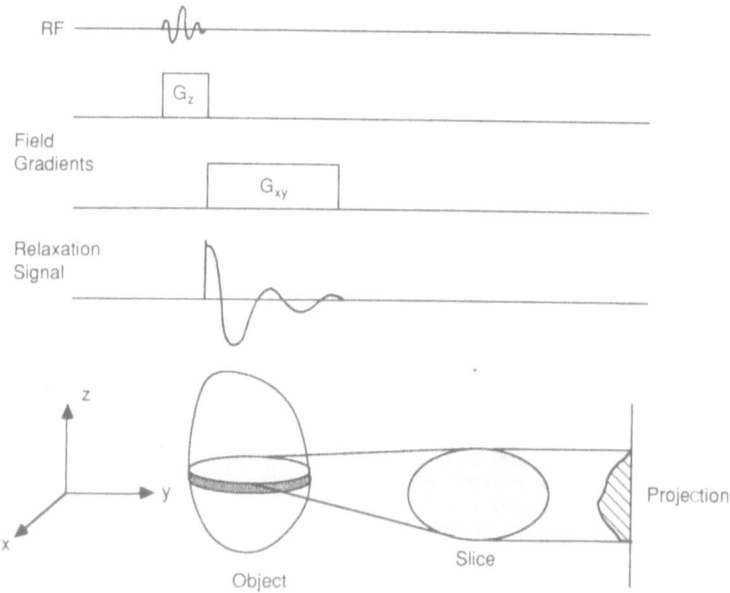

Fig. 2.26. Measurement sequence for two-dimensional slice scanning

puterized tomography and the object can be reconstructed using the same techniques (see Sect. 2.4). By using different sequences of RF pulse and subsequent measurement, the decay curves and thus the spatial distribution of T_1 or T_2 can be constructed.

The medical NMR imaging system consists of a large magnet (often a superconducting magnet) surrounding the patient, who lies along the axis of the magnet (Fig. 2.27). The magnet produces a uniform magnetic field over a large volume. A gradient coil system (Fig. 2.28) induces gradients $G_z = \partial B_z/\partial z$, $G_x = \partial B_x/\partial x$, $G_y = \partial B_y/\partial y$ on the large uniform field to enable spatial information to be collected. The RF system consists of a transmitting antenna and a receiving antenna, usually either solenoidal or saddle shaped. Figure 2.27 shows a saddle-shaped coil, which gives easy access to the patient. In some cases the same coil is used to irradiate and to receive the NMR signal, or different coils can be used. More sensitive measurements are possible if a surface probe is employed. A computer controls data acquisition; it controls the magnetic gradient coils and the RF pulse sequencing and collects the measured data, which it then processes and displays.

2.5.2 Image Quality

The image quality is defined by spatial resolution, measurement time, and contrast resolution. The interdependence of these parameters can be seen by examining some of the factors which affect the signal to noise ratio.

$$S/N \approx K \frac{B_0^{3/4}}{T_b^{1/2}} V_0 \chi_0 (mt)^{1/2} \qquad (2.35)$$

where K is a system sensitivity coefficient, T_b the temperature of the RF coil, V_0 the voxel size, χ the

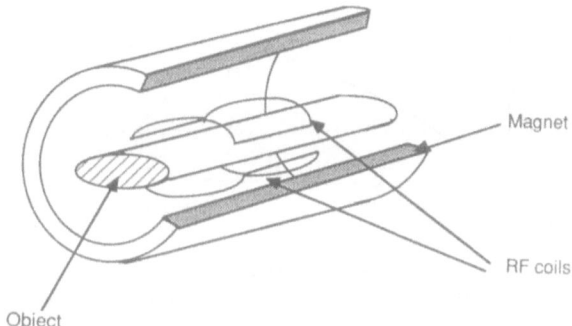

Fig. 2.27. Position of patient (object) in relation to main magnet and RF coils

magnetic susceptibility, m the number of cycles, t the acquisition time, and B_0 the principal magnetic field strength.

The spatial resolution or voxel size is a function of the steepness of the gradient field applied, and the bandwidth of the RF excitation and detection (see Fig. 2.25). From Eq. 2.35, it can be seen that as the voxel size is decreased, the signal to noise ratio decreases as a linear function. To retain the same signal to noise, either the measurement time (mt) or the field strength B_0 must be increased, or the system sensitivity coefficient K increased; K depends on the quality of the magnetic and RF components and hence on the cost of the machine. More specifically, an increase in the signal to noise and thus the contrast resolution by a factor of 2 requires either an increase in measurement time by a factor of 4, a decrease in spatial resolution by a factor of 2, an increase in the principal field by a factor of 1.58, or a doubling of system sensitivity.

In reality, the relationship between signal to noise ratio and field strength is more complex than Eq. 2.35 suggests (see Fig. 2.29). The signal increases more than linearly with field strength; however, as field strength increases, so do both the magnitude of inhomogeneities in the field and the magnitude of the chemical shift. Thus, since the field gradient must be

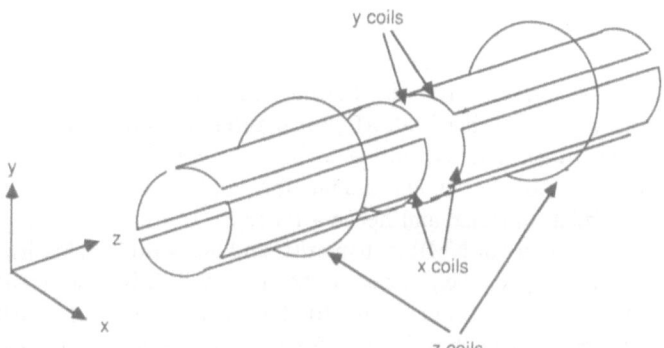

Fig. 2.28. Gradient soil system

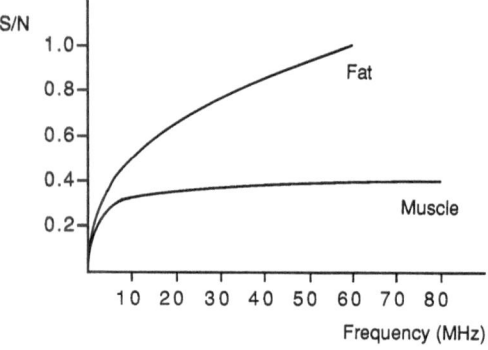

Fig. 2.29. Signal to noise (S/N) ratio versus RF frequency for different tissue types (partial saturation sequence repetition time = 500 mS). (By courtesy of Siemens Medical Engineering Group)

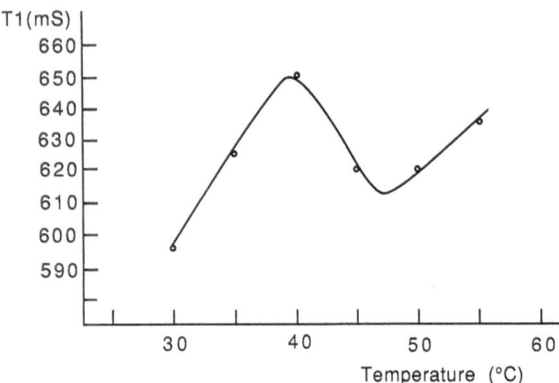

Fig. 2.30. Temperature dependence of T_1 relaxation time for excised muscle tissue. (Data from Lewa and Majewska [76])

large compared to these inhomogeneities, as field strength is increased, field gradients must also be increased. If the gradient is doubled, then the bandwidth per pixel is also doubled, increasing the image noise by a factor $\sqrt{2}$. This means that the S/N versus B_0 curve flattens out at higher field strengths. In addition, T_1 relaxation time increases with increasing field. As a result the acquisition time must be increased to retain the same S/N ratio. This relationship between T_1 and field depends on tissue type, and consequently, so does the relationship between S/N and field. Finally, since the Larmor frequency increases with increasing field, RF propagation in tissue must be considered. Operating at higher frequency leads to a greater signal attenuation and phase displacement in tissue and hence a lower S/N ratio. Since many types of NMR system exist, and many different measurement sequences can be employed, it is difficult to give definite figures for typical imaging quality parameters. Spatial resolution of less than

1 mm can be achieved with most machines for smaller areas of investigation, although values of $1-5$ mm are more typical for larger body sections. Measurement times depend on the parameter to be investigated. For the T_1 parameter, data acquisition times of $5-10$ min are typical when two different repetition times are used to estimate the T_1 curve and two scans are averaged to increase signal to noise. Fast data acquisition algorithms exist which can reduce measurement times to a few tens of seconds for high magnetic field systems, with degradation of signal to noise ratio. Contrast resolution for T_1 is typically in the range $1\% - 5\%$.

2.5.3 Temperature Dependence of NMR Imaging Parameters

The parameter which has been used most successfully for obtaining temperature data is the spin-lattice relaxation time T_1 of water protons, for which the primary relaxation mechanism is interaction with the thermal motions of the surrounding molecules. Other parameters, for example the chemical shift of water protons, exhibit a temperature dependence and have been suggested for possible use in hyperthermia monitoring [70]. The fast-exchange two-state (FETS) model for the proton relaxation of water gives the T_1 temperature dependence (above a certain minimum temperature) approximately as

$$T_1 \approx T_1^\infty \, e^{-E_a/kT} \qquad (2.36)$$

where T_1^∞ is the value of T_1 for a temperature tending to infinity, E_a is the activation energy of the relaxation process, and k is Boltzmann's constant. Figure 2.30 shows the temperature dependence of T_1 for excised muscle tissue. It is important to note that Lewa and Majewska [76] found that T_1 decreases irreversibly for the range $T > T_{T1max}$, the magnitude of these changes depending on the heating temperature in the range $T_{T1max} < T < T_{T1min}$. The shape of this T_1 versus T curve for tissue is strongly dependent on the heating time of the samples. According to the model mentioned above, this irreversibility can be explained either by an increase in the concentration of the fraction of water in tissue which is bound to proteins, or by an acceleration of the spin-lattice relaxation process, or by the combined effect of these two factors. This irreversibility, if it is also found for in vivo experiments, will have serious consequences for the use of NMR as a hyperthermia monitoring technique, since T_{T1max} occurs for most tissues at around 40 °C.

2.5.4 Some Results

Experiments have been conducted to measure the temperature dependence of T_1 in water and blood samples [103], excised tissue samples [36, 76], and in vivo [37]. The results of Lewa and Majewska are presented in Fig. 2.30. Similar experiments on excised tissue by Dickinson et al. do not show the irreversible change in T_1, but show that T_1 increases linearly with temperature in the range 10°–60 °C. The difference between the results appears to be due to the fact that the heating times for Dickinson's experiment [36] were much shorter than those for Lewa and Majewska [76].

In vivo temperature monitoring has been performed in rabbits and in the lower leg and lower arm of human volunteers. The in vivo results confirm the linear dependence of T_1 on temperature for short periods of heating (Fig. 2.31). The analysis of results included a subtraction procedure, taking the difference between the image after heating and that before heating. This was done because T_1 varies more with tissue type than with temperature. Results from the lower leg showed a T_1 temperature dependence of 2%/°C for given regions of interest of the scan compared with temperature measurements by implanted thermocouples. For a magnetic field of 0.15 T and an acquisition time of 72 s per scan, a temperature resolution of 1°–2 °C was achieved with a spatial resolution around 4 mm. Heating was performed by a capacitive system placed inside the bore of the ma-

chine. In the case of the lower arm, heating was performed by circulating hot water through a tube coiled around the arm and thermometry consisted of thermocouples on the skin surface. A T_1 temperature dependence of 1%–3%/°C was found with a temperature resolution of about 1 °C for a measurement time of 150 s and a spatial resolution of 2–3 mm. The better accuracy is attributable to three factors: the longer scan time, the use of a surface antenna, and a stronger magnetic field of 0.35 T. Figure 2.32 shows the difference between two images of the forearm at skin temperatures of 40 °C and of 37 °C, illustrating the increase in superficial T_1 values.

2.5.5 Discussion

Nuclear magnetic resonance imaging shows some very promising features as a method for noninvasive thermometry. The technique gives very good spatial resolution, although this could be compromised by averaging procedures in order to improve the temperature resolution or reduce scan times. Another enormous advantage is the technique's three-dimensional imaging ability, giving more flexibility to the monitoring procedure. The present temperature resolution of 1°–2 °C may be adequate for hyperthermia monitoring. Scan times of more than 1 min could preclude the use of this technique in real-time operation, since for in vivo experiments it was found necessary to gate the

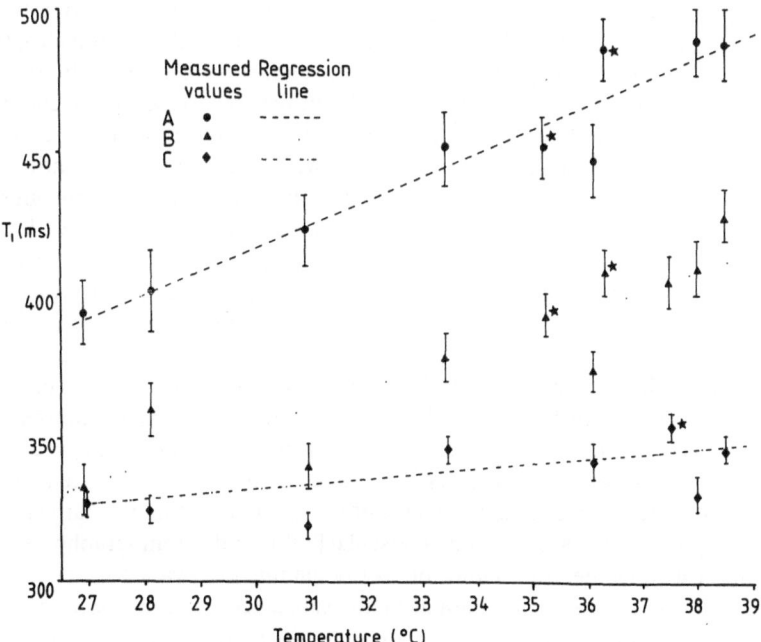

Fig. 2.31. The temperature dependence of T_1 of human calf muscle measured in vivo. [●, ▲] in the heated field; [♦] outside the heated field. Measured on heating except [∗] measured on cooling. (By courtesy of R. J. Dickinson)

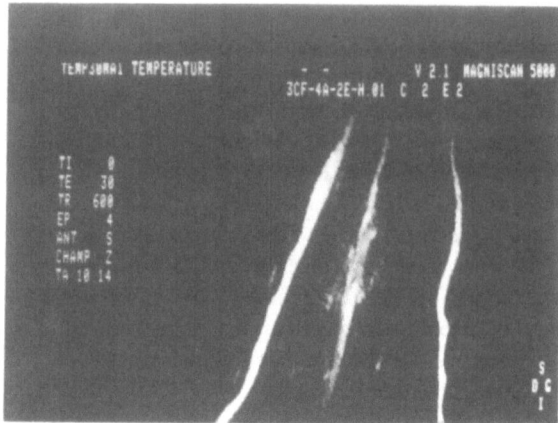

Fig. 2.32. Difference between two NMR images of the forearm before (skin temperature 37°C) and after (skin temperature 40°C) heating. The heating fluid, which circulates in a tube, is visible at the forearm periphery. (By courtesy of F. Micheron)

heating power off during data acquisition, which may lead to inefficient heating sequences. However, since the technique has only recently been applied to temperature measurement, it can be imagined that these parameters could be improved upon in the future.

Additional difficulties are similar to those found for X-ray tomodensitometry, and indeed, to some extent for all noninvasive thermometry techniques. First, compatibility between the NMR scanner and some heating systems may be difficult to achieve due to the lack of space within the bore of the magnet and due to possible interaction of hyperthermia equipment with magnetic or RF fields. Dickinson et al. solved these problems by using a compact method of administering the heating field and by interrupting the heating during data acquisition. Second, the temperature dependence of T_1 is different in different tissues, requiring a priori knowledge of tissue composition for calibration. However, the available knowledge of tissue type from NMR scanning means that this may not be a large problem. Perhaps more important is the dependence of the image on other physical parameters which may change during hyperthermia, e.g., blood flow. Since the slope of T_1 versus temperature in the in vivo experiments was similar to that found in vitro [37], it could appear at first that these other factors have no influence on the image. However, more experimental data are needed to confirm these initial findings since NMR images are more sensitive to differences between tissue compositions than to tissue temperature. Knüttel and Juretschke [70] found that T_1 as a temperature indicator will be hampered by its high sensitivity to paramagnetic substances, e.g., O_2 or other radicals. Since these quantities are

not uniformly distributed throughout the body, a varying influence is to be expected.

One of the most serious question marks over the use of NMR imaging for hyperthermia monitoring is the problem of the irreversible change in T_1 which has been demonstrated when tissue samples are heated above 40°C for long periods. Although this effect was not evident in in vivo results, it is not clear whether this is due to short measurement times or whether this effect would appear during the course of a normal hyperthermia treatment of around 1 h duration.

Finally, the high cost of NMR systems at present probably precludes their use for routinely monitoring hyperthermia treatments in most hyperthermia clinics. If system costs could be significantly reduced, NMR would show promise as a viable means of noninvasive temperature monitoring.

2.6 Imaging of Dielectric Properties

2.6.1 Presentation

As indicated in Sect. 2.2, electromagnetic waves are extensively used in medical imaging. Formally, the basic phenomena can be described by means of the same equations – namely, Maxwell's equations – over all the spectrum extending from low frequencies to the visible range. Nevertheless, the mode of propagation and the image content strongly depend on the specific properties of living tissues and on the dimension to wavelength ratio of the considered structures. Accordingly, the tools to be used to describe the interaction phenomena must be suitably chosen in order to simplify, as much as possible, the mechanism of image formation. For this purpose, it is suggested that the electromagnetic spectrum be divided into at least three parts within each of which there exists some unity in the description of the propagation phenomena. In the first part, extending from DC currents to a few tens of megahertz, it is possible to assume that, roughly speaking, the cross-section body dimensions are small as compared to the wavelength. Quasi-static analysis can be efficiently used to determine current and voltage distributions induced by electrodes or coils located around the body under investigation. Simple impedance measurements or conductivity tomography – as well as NMR imaging – are relevant to such an analysis.

In the second part, the propagation aspect of the interaction predominates. The considered wavelengths are of the order of or smaller than the cross-sectional body dimensions and diffraction and ray propagation are most appropriate. In all cases, the approaches are efficiently oriented toward an optical vision of the interaction phenomena even if, in some cases, a drastically heavier formalism has to be used in order to obtain the desired degree of accuracy. Microwave passive and active imaging are typical examples of such complicated situations. The case of infrared radiations, although simpler, is unfortunately of limited usefulness due to their strong attenuation in living tissues. The visible range is now beginning to be considered for tomographic imaging, with the additional problem of light diffusion.

It is necessary to reach ionizing radiations, such as X-rays, to recover quasi-pure linear propagation. This domain has already been considered in Sect. 2.4.

This section is concerned with the first two spectrum parts, which allow imaging of the dielectric properties of tissues. Dielectric properties are conveniently described by macroscopic constants such as the dielectric constant or conductivity. The specific aspects of three different dielectric imaging approaches will be discussed in the following subsections. The basic equations required to describe wave-tissue interactions and the fundamental dielectric properties of tissues will be briefly addressed. The first two approaches described are electrical impedance tomography and active microwave imaging. In both cases, the principle and equipment will be described. From analysis of available results and from their extrapolation, the advantages and the limitations of these two techniques will be presented. Finally, inverse scattering techniques are more prospectively discussed as a possible generalization of the two aforementioned approaches.

2.6.2 Fundamentals

Maxwell's equations are known to provide an adequate macroscopic description of electric $\mathbf{E}$ and magnetic $\mathbf{H}$ vector fields as well as currents and charge distributions inside matter. A very general approach consists of decomposing the total field $\mathbf{E_t}$, $\mathbf{H_t}$ inside a given medium as the summation of two terms:

$$\mathbf{E_t} = \mathbf{E_a} + \mathbf{E_s}$$
$$\mathbf{H_t} = \mathbf{H_a} + \mathbf{H_s} \ . \tag{2.37}$$

The first one is expected to take into account the applied field resulting from the excitation devices, electrodes, or antennas. This applied field $\mathbf{E_a}$, $\mathbf{H_a}$ is

assumed to be known and usually can be seen as the field which would exist in homogeneous medium where the body under investigation will be immersed (air, water, etc.).

The introduction of the body perturbs this initial applied field distribution. The perturbation is accounted for by a scattered field $\mathbf{E_s}$, $\mathbf{H_s}$ which depends on the body shape, dimensions, and electromagnetic properties. The latter can be described, at a given frequency, in terms of complex permittivity ε and complex permeability μ. The term "complex" indicates that the media can exhibit losses. As shown later, these properties change significantly with frequency.

It is conceptually very useful, and theoretically grounded [e.g., 58], to consider that the second field $\mathbf{E_s}$, $\mathbf{H_s}$, which is unknown, results from the radiation of charges and currents induced by the applied field in the body (Fig. 2.33). Let the body be nonmagnetic, then the induced equivalent currents – which implicitly include charges by means of the so-called continuity equation – are only electric currents and they can be written as follows:

$$\mathbf{J}(\mathbf{x}) = j\,\omega\{\varepsilon(\mathbf{x}) - \varepsilon_a\}\mathbf{E_t}(\mathbf{x}) \tag{2.38}$$

where ε_a designates the complex permittivity of the ambient medium.

It is then worth noting that these induced currents depend, through the first term, on the permittivity contrast between local properties of the body and those of the medium where it is immersed, while the second term depends on the total field existing at the considered point $\mathbf{x}$.

The current $\mathbf{J}$ includes conductivity and the dielectric constant of the body. More precisely, the complex permittivity can be separated according to its real and imaginary parts:

$$\varepsilon(\mathbf{x}) = \varepsilon'(\mathbf{x}) - j\,\varepsilon''(\mathbf{x}) \ . \tag{2.39}$$

The imaginary part is related to loss mechanisms resulting from ohmic conduction or dielectric resonance. In the first case, the conductivity σ can be explicitly introduced as:

$$\varepsilon''(\mathbf{x}) = \sigma(\mathbf{x})/(\omega \cdot \varepsilon_0) \tag{2.40}$$

whereas the relative dielectric constant ε_r is such that:

$$\varepsilon'(\mathbf{x}) = \varepsilon_0 \varepsilon_r(\mathbf{x}) \ . \tag{2.41}$$

The loss angle Θ is often introduced by its tangent such that:

$$\mathrm{tg}\,\Theta = \varepsilon''/\varepsilon' \ . \tag{2.42}$$

It is evident from previous equations that, at low frequencies, the induced currents reduce to the conduction current satisfying Ohm's law:

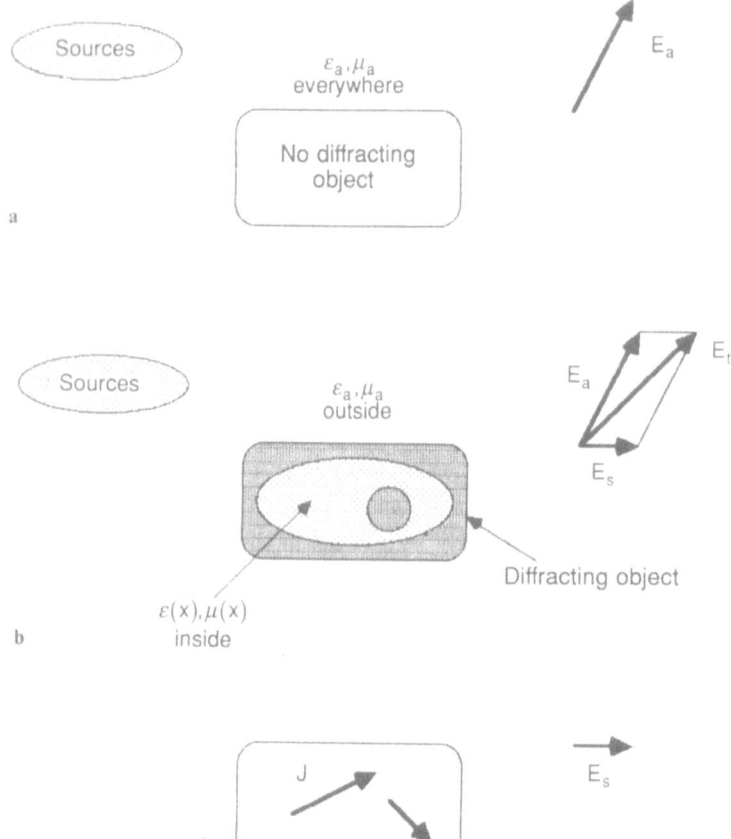

Fig. 2.33a–c. On the equivalent current concept: the perturbation $\mathbf{E_s}$ resulting from the introduction of an inhomogeneous body in an applied electromagnetic field $\mathbf{E_a}$ can be considered as radiated from equivalent currents $\mathbf{J}$ radiation in a homogeneous medium

$$\mathbf{J}(\mathbf{x}) = \sigma(\mathbf{x})\,\mathbf{E_t}(\mathbf{x}) \tag{2.43}$$

At such frequencies were conduction currents predominate, it is convenient to consider that the electric field derives from a potential function V, such that:

$$\mathbf{E_t}(\mathbf{x}) = -\operatorname{grad} V_t(\mathbf{x}) \; . \tag{2.44}$$

The field decomposition can be applied to the potential V as well. So, the total potential V_t can be written as:

$$V_t = V_a + V_p \tag{2.45}$$

where the first term corresponds to the potential distribution in a homogeneous medium, and the second results from the perturbation introduced by some inhomogeneity. For terminological purposes, let us just note that in the low-frequency range, the term "scattering" is no longer used and the term "perturbation" is employed. The previous decomposition of the potential V_t will be introduced later in the section on electrical impedance tomography.

Knowledge of ε' and ε'' allows calculation of the complex propagation constant γ whose real and imaginary parts are the attenuation constant α and the phase constant β [98]:

$$\alpha = \frac{2\pi\sqrt{\varepsilon'}}{\lambda_0}\{(1+\operatorname{tg}^2\Theta)^{1/2}-1\}^{1/2} \; , \tag{2.46}$$

$$\beta = \frac{2\pi\sqrt{\varepsilon'}}{\lambda_0}\{(1+\operatorname{tg}^2\Theta)^{1/2}+1\}^{1/2} \tag{2.47}$$

where λ_0 is the free-space wavelength.

As already indicated in Sect. 2.3, the penetration depth $\delta = 1/\alpha$ is the distance over which the electromagnetic field is reduced by $1/e$ in the case of plane wave propagation.

Both σ and ε_r are dependent on some physical or physiological properties of tissues. Basically, the purpose of dielectric imaging is to reproduce their spatial distribution in the observed volume. In the particular cases where it is really possible to retrieve the complex permittivity distributions, the corresponding imaging can be said to be *quantitative* (see Sect. 2.2). More generally, only equivalent currents are directly accessible through a linear inversion process, in such a way that only *qualitative* imaging is easily achieved. "Qualitative" means that the spatial distribution of the complex permittivity is weighted by the total field variations. But, when the total field is constant or slowly varying, or when the observed tissues present very high contrasts of complex permittivity, the equivalent current distribution is generally representative of the anatomical structures.

The imaging problem consists of retrieving the equivalent current distribution from the measurement of the scattered field E_s, H_s over a suitable area external to the object, for some given known incident fields E_a, H_a. In order to solve this problem, knowledge of the basic relationship between equivalent currents and scattered fields is required. These fields can be conveniently expressed as an integral over the equivalent current support [e.g. 93]:

$$Es(x) = \frac{1}{j\omega\varepsilon_a}\{k_a^2 + \text{grad div}\}\iiint G_a(x,x')J(x')d^3x' \quad (2.48)$$

$$H_s(x) = \text{rot}\iiint G_a(x,x')J(x')d^3x'$$

with:

$$Ga(x,x') = -\frac{\exp(-jk_a|x-x'|)}{4\pi|x-x'|} \quad (2.49)$$

$$k_a = \omega(\varepsilon_a\mu_a)^{1/2} \ .$$

The direct diffraction problem consists in deducing J from the knowledge of the applied field and the body structure. For a known body, these equations can be used to derive the applied field producing a given current distribution or, equivalently, a specified power deposition. For this reason, as discussed on p. 100 they constitute the foundation of optimization of hyperthermia treatments.

Conversely, the imaging process consists of inverting these equations in order to obtain the equivalent cur-

rent distribution as a function of the scattered field. Different techniques can be used according mainly to the frequency range under consideration and to the selected experimental setup geometry.

2.6.3 Salient Features of Dielectric Properties of Living Tissues

The dielectric characterization of living tissues has been considered for a long time [e.g. 21, 47]. Extensive measurements have been performed, mostly in vitro but sometimes in vivo, which enable some salient features to be pointed out.

First, the complex permittivity depends on the tissue nature, or on its constitution. It is convenient to distinguish between high water content tissues, such as muscle, liver, and kidney, and low water content tissues, such as bone and fat. In the first case, water impresses its major characteristics on the tissue: a high dielectric constant and significant conductivity. As a result, the wavelength is significantly smaller in tissues than in air, and losses cannot be neglected.

In the second case, the dielectric constant is much lower and conductivity very low, resulting in larger wavelengths and reduced losses. Table 2.4 gives some significant results illustrating these tendencies. The direct current (DC) and microwave cases (f = 3 GHz) have been considered for electrical impedance and microwave tomography, respectively.

Secondly, for a given tissue, the complex permittivity strongly depends on the wavelength, or equivalently on the frequency. Water provides an example of a simple medium where the influence of the frequency can be seen. Water is a simple medium in the sense that only water molecules are present. These molecules are responsible for a resonance region in the microwave domain. Below the corresponding resonance frequency, around 20 GHz, the dielectric constant decreases and the losses increase. Living tissues have more complicated structures and are composed from the arrangement of different molecules. The variations of the complex permittivity are more complicated, resulting from independent or coupled resonances of these molecules or arrangement of them.

Thirdly, for a given tissue at a given frequency, many other factors are able to influence the complex permittivity. Among these factors, water content is probably one of the most important. But temperature changes are also responsible for small complex permittivity changes. Temperature coefficients of a few percent per degree Celsius seem to be typical. Figure 2.34 shows the temperature dependence of some

Table 2.4. Typical direct current conductivity and microwave complex permittivity of water and different biological media

Medium	DC (f = 0 Hz) σ(S/m)	Microwave (f = 3.00 GHz)			
		ε_r	σ(S/m)	λ(cm)	α(dB/cm)
Water		77	1.57	1.16	2.9
Muscle	0.19	46	2	1.5	4.8
Blood	0.66	52	2.7	1.38	6.1
Fat	0.05	5.5	0.13	4.3	0.9
Bone	0.006	8	0.22	3.5	1.3

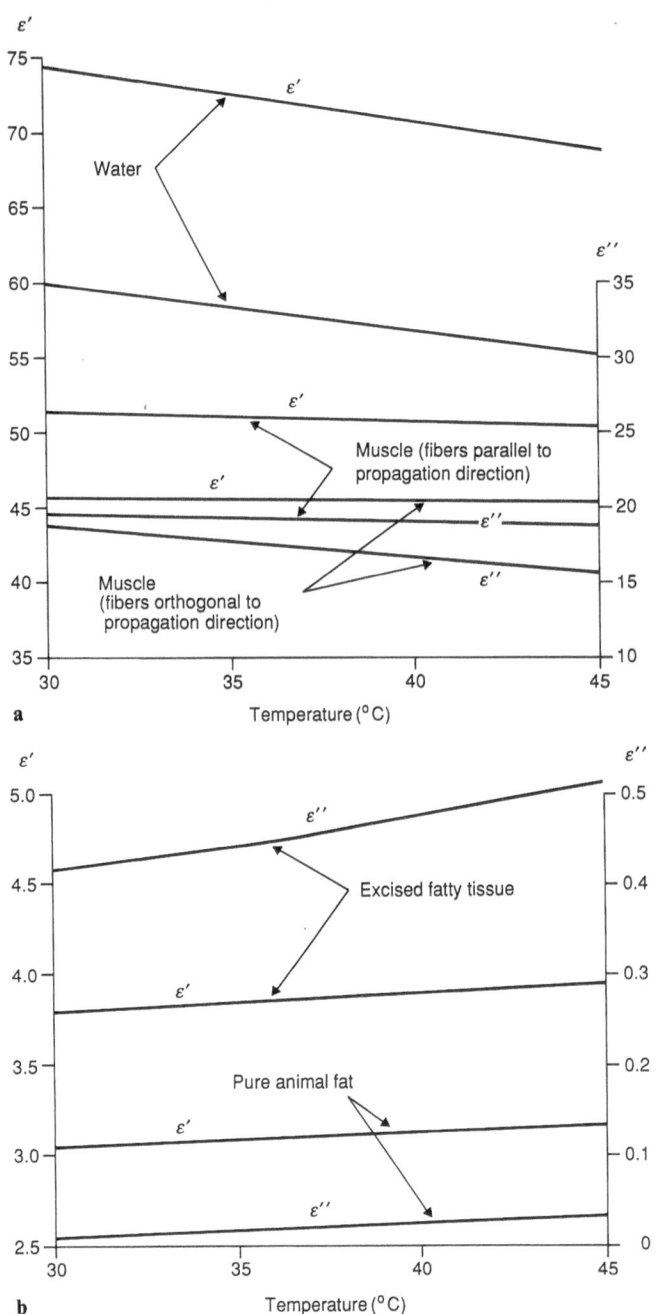

Fig. 2.34a, b. Temperature dependence of the dielectric properties of different media around 3 GHz (Guerquin-Kern et al. [50])

tissues obtained via in vitro measurements in the microwave domain. It is worth noting that such an order of magnitude for the temperature coefficient has been obtained from measurements on excised tissues, excluding the influence of blood circulation. In the hyperthermia context, care must be taken due to the fact that, beside the previous intrinsic permittivity variations resulting from temperature changes, tissue modification under thermoregulatory mechanisms may also introduce permittivity changes.

As is now well known, temperature changes can induce significant variations of the blood flow rate or perfusion. Measurements have been performed by de Lateur et al. [33] on human thighs heated by 915 MHz radiation. Using an Xe 133 clearance technique, they showed that the blood flow rate was increased from, approximately, 4 to 36 ml/min/100 g of tissue while the temperature was increased from 36 °C to 45 °C at 2.1 cm depth. Other measurements have been achieved by means of thermal clearance techniques [84].

More recently, Milligan et al. [80] developed a simple thermal model to predict the blood flow rate change. The model was validated from experiments on canine limbs by means of a thermal washout technique. With a correlation coefficient of 0.991, the blood flow rate peak value B fits with the following least squares approximation:

$$B = 3.15 \exp[372(T-37)] \qquad (2.50)$$

where B is given in ml/min/100 g of tissue and T is the treatment temperature in °C. Starting from an initial blood flow rate of a few units, peak values of 38−138 ml/min/100 g of tissue were observed for treatment temperatures varying between 43 °C and 47 °C.

These two examples demonstrate the drastic blood flow change during hyperthermia treatments. On the other hand, careful experiments have been conducted by Burdette [20] demonstrating the correlation between dielectric properties and the blood flow rate. The complex permittivity was measured by means of coaxial probes at 2.45 GHz on canine gastrocnemius and soleus muscle (four animals, multiple locations), completely isolated from all blood vessels, except the caudal femoral artery whose blood flow was externally controlled. From the results it appeared that the complex permittivity of well perfused tissues is a predictable function of blood flow through a linear relationship. The corelation coefficients were 0.931 and 0.970 for the dielectric constant and the conductivity, respectively. Blood flow change from 0 to 15 ml/min/100 g of tissue produced correlative variation of the dielectric constant of 5 units. Tissue volume and weight changes were directly proportional to the blood flow rate, indicating that the dielectric properties largely result from bulk flow in the muscle tissue. However, tissue volume changes did not completely account for the close correlation between dielectric properties and blood flow. According to careful experiments on isolated vessels, axial streaming seems to responsible for these additional effects.

As a consequence, the temperature coefficient of a dielectric image is dependent on the space location through the tissue nature. But, at a given point, it also varies with time: it is not the same during the first minutes of the hyperthermia treatment (linear temperature increase) as after the moment the thermoregulatory process comes into effect (temperature plateau). Such a situation explains the difficulties in effectively realizing absolute thermometry. But even if absolute calibration remains impossible, the variations of the images with temperature changes could yield new and interesting information on the heating process.

2.6.4 Electrical Impedance Tomography

2.6.4.1 Presentation

Electrical impedance tomography (EIT), or applied potential tomography (APT), is a relatively new technique which nevertheless already has many possible clinical applications. These clinical applications involve the measurement of some property within the body which causes a corresponding change in resistivity, for example, the visualization of blood perfusion in the heart and lungs, respiratory function, stomach emptying, etc. [18, 43, 59].

Recently, a lot of interest has been shown in the technique as a possible means of monitoring hyperthermia treatment. One of the main points of interest with such a system is that, operating at low frequencies, it can be constructed very inexpensively using well-known technology and mostly mass-produced electronic components.

2.6.4.2 Fundamentals

Impedance tomography operates at frequencies of a few tens to a few hundreds of kilohertz. At such low frequencies, the wavelength is much larger than the dimensions of the human body so that the quasi-static condition operates; that is, the analysis can assume d.c. conditions. Current and voltage on the surface of a volume are measured to determine the distribution of resistivity (or conductivity) within that volume.

The distribution of electrical potential V within an inhomogeneous anisotropic conducting medium through which a steady current is flowing is given by Poisson's equation

$$\operatorname{div}[C \operatorname{grad} V] = 0 \qquad (2.51)$$

where C is the space-variant conductivity tensor. If appropriate boundary conditions are specified, such as V or the gradient V at certain points in space, unique solutions can be found for Eq. 2.51. If the conductivity is uniform and isotropic, Eq. 2.51 reduces to

$$\nabla^2 V = 0 \qquad (2.52)$$

which is Laplace's equation. If the medium is isotropic but nonuniform

$$\sigma \cdot \nabla^2 V + \operatorname{grad} \sigma \cdot \operatorname{grad} V = 0 \qquad (2.53)$$

where σ is the conductivity.

Replacing σ by the logarithmic resistivity $R = -\ln \sigma$

$$\nabla^2 V - \operatorname{grad} R \cdot \operatorname{grad} V = 0 \qquad (2.54)$$

If a given current distribution is applied to the boundary, the voltage distribution on the boundary depends on the resistivity distribution of the volume. For a series of independent current distributions, it is possible to approximately reconstruct the resistivity distribution from boundary measurements of potential.

Three types of solution are possible for the above equations. Analytical methods attempt to find a general solution and then apply boundary conditions to solve particular cases. This type of solution is only possible in very simple geometries. Numerical methods divide the conductivity distribution into a set of discrete elements and then apply finite element methods to solve the resulting equations. Alternatively, approximate solutions can be formulated by means of linearizing Eq. 2.54 upon imposition of certain assumptions. Writing V as

$$V = V_p + V_a \qquad (2.55)$$

where V_p is a perturbation on the potential distribution V_a, which is the potential distribution for a uniform conductivity distribution, Eq. 2.54 can be written

$$\nabla^2 V_p = \text{grad } V_a \cdot \text{grad } R + \text{grad } V_p \cdot \text{grad } R \qquad (2.56)$$

and if conductivity variations from uniformity are small, $\text{grad } V_p \ll \text{grad } V_a$ and

$$\nabla^2 V_p = \text{grad } V_a \cdot \text{grad } R \ . \qquad (2.57)$$

The potential distribution is now a linear function of the resistivity distribution and so can be solved by linear methods.

One such approximate linear reconstruction method has been described by Barber et al. [5] and Barber and Seagar [6], allowing fast reconstruction of resistivity change distributions by employing a technique well known in X-ray tomography, that of filtered back-projection. Current is injected between two electrodes and the potential is measured between each of the other electrode pairs placed in a circular geometry. Figure 2.35 shows lines of constant potential which end on the electrodes. The reconstruction method consists in measuring the potential difference between two electrodes and comparing this difference to that measured for the reference data set. If there has been a change in potential difference, the resistivity of the region between the constant potential lines is altered in proportion to this potential difference change. In other words, the change in potential difference is back projected into the region between isopotentials. An image is built up as the current drive is applied to different pairs of electrodes. Image processing is performed by means of applying weighting factors to the back projections to homogenize the spatial response

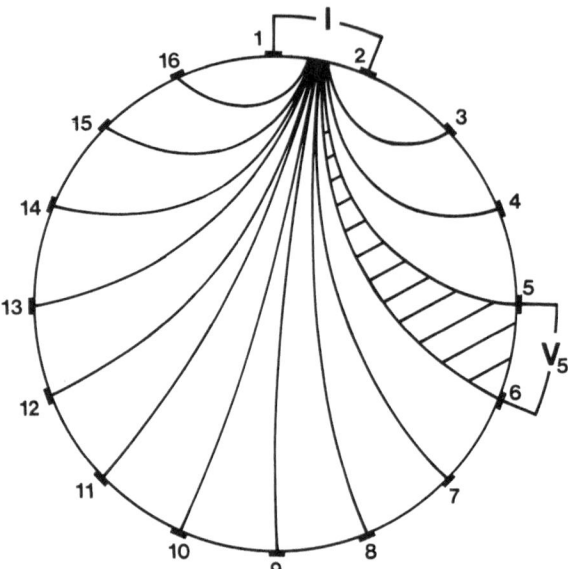

Fig. 2.35. Lines of constant potential which end on electrode locations in a uniform phantom for a current drive I between electrodes 1 and 2. In the Barber-Brown reconstruction method, a change in the potential difference δV_5 is back-projected into the *shaded region*. (By courtesy of B.H. Brown)

of the reconstruction. This method implicitly assumes a circular boundary. If this is not the case, for example if the electrodes are placed around the torso, errors will certainly result. Additional errors occur due to the linearizing assumption. Many different reconstruction methods have been applied to EIT [e.g. 97]. Yorkey et al. [99] have compared different reconstruction techniques and conclude that a numerical algorithm, that of Newton-Raphson, gives the best performance for simulated data.

In practice, measurements are performed by placing electrodes in contact with the skin in a ring around the part of the body to be measured. The electrodes employed for EIT are often the same as those used in electrocardiography. Many different measurement schemes are possible, depending on the object to be measured. In one measurement configuration, constant current drive is injected between two adjacent electrodes and the potential difference between all remaining pairs of electrodes is measured. The current drive is then moved to the next set of electrodes, and so on. Figure 2.36 shows a typical data collection system.

2.6.4.3 Image Quality: Basic Limitations

Definition of the imaging ability of a system requires consideration of the sensitivity (or conductivity

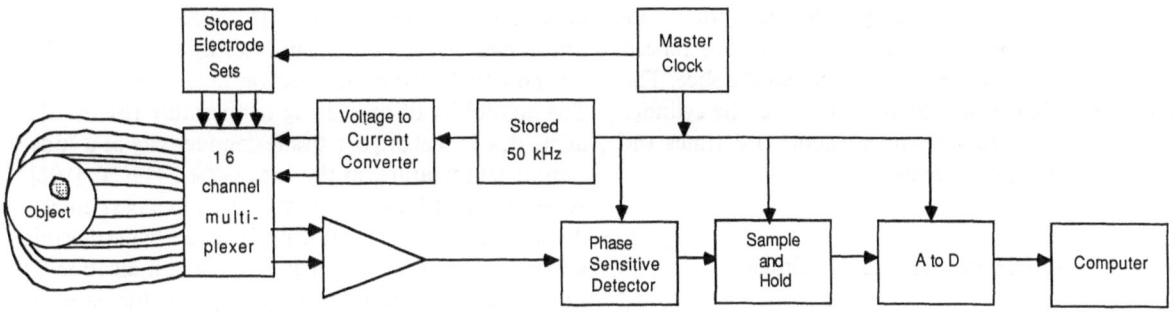

Fig. 2.36. A typical data collection system for EIT. (By courtesy of B. H. Brown)

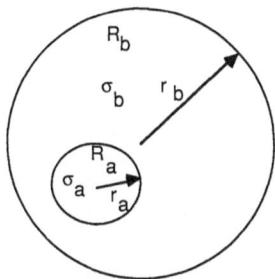

Fig. 2.37. Conductive region R_a with radius r_a and conductivity σ_a containing smaller conductive region R_b with radius r_b and conductivity σ_b

resolution), the spatial resolution, and the data acquisition time. For impedance imaging, these are related in a complex manner. Some of the interrelationships and limitations on these quantities can be derived theoretically; others are a function of hardware and software design. A thorough analysis of these factors has been carried out by Seagar et al. [94] and Seagar and Brown [95], and some of the results will be given here.

With reference to Fig. 2.37, consider a circular conductive region R_b having radius r_b and conductivity σ_b which contains a smaller circular region R_a with radius r_a and conductivity σ_a. The spatial resolution is defined as the smallest region R_a in which the conductivity can be independently determined, and is quantified here by r_a/r_b. The conductivity resolution is the smallest range into which the conductivity of a region can be isolated and is given as $d\alpha/\alpha$ where α is the conductivity contrast and is given by σ_a/σ_b. If dV/V is taken to be the noise in the peripheral measurement of potential, the following interrelationships can be defined:

1. Spatial resolution-conductivity resolution; degrading conductivity resolution by a factor K balances improving spatial resolution by $\sqrt{K}$.
2. Spatial resolution-noise; improving noise by K balances improving spatial resolution by $\sqrt{K}$.
3. Conductivity resolution-noise; improving noise by K balances improving conductivity resolution by K.
4. Spatial resolution-contrast; extending contrast by K balances degrading spatial resolution by $\sqrt{K}$.
5. Conductivity resolution-contrast; extending contrast by K balances degrading conductivity resolution by K.
6. Noise-contrast; extending contrast by K balances improving noise by K.

As an example, if the noise is improved by a factor of 10, this could give an improvement in spatial resolution by a factor of 3.2 or an improvement in conductivity resolution by a factor of 10. Spatial resolution is evidently the least sensitive of these factors to changes in the others.

In addition to these theoretical relationships, system design and clinical applicability enforce additional compromises and limitations. Increasing the number of electrodes to improve spatial resolution leads to a proportional increase in noise. This then leads to a decrease in spatial resolution which may wipe out the advantage gained. Measurement time and signal to noise (S/N) ratio are strongly related. Shortening measurement time decreases the accuracy of measurement of potential. In order to improve the S/N ratio, common mode feedback is often used, making parallel data collection impossible and thus greatly lengthening data collection time. Another means of increasing S/N is to average over a number of data sets. This method also has the advantage of averaging out noise due to patient movement etc., but again leads to an increase in measurement time.

Typical performance figures for a 16-electrode system give a spatial resolution of around 10% of the diameter of the circle around which the electrodes are placed, a conductivity resolution of around 3% for volumetric changes, and a data collection time per frame of 0.1 s. The slice thickness depends on the current path between electrodes in the plane perpendicular to the plane containing the electrodes. The tomographic slice for the case of electrodes placed around a uniform cylinder varies across the slice. The sensitivity falls off to −20 dB at 0.5 times the cylinder radius at the periphery and at about 1.0 times the cylinder radius in the center.

2.6.4.4 Use in Hyperthermia Monitoring

Electrical impedance tomography for monitoring hyperthermia has been developed very recently. Some initial phantom studies and in vivo experiments will be described in this section. The assessment of EIT for clinical hyperthermia monitoring has begun [31] but no conclusive results are yet available.

The possibility of measuring temperature rise by EIT arises due to a change in tissue conductivity as a function of temperature, in the range 1%−4%/°C [102]. An example of how EIT may be used in a particularly difficult case, the monitoring of deep hyperthermia within the chest cavity, will give some indication of the system specifications necessary. In this case the electrodes can be placed around the torso to give a tomographic slice of the chest. It is possible to estimate the sensitivity necessary in recording peripheral

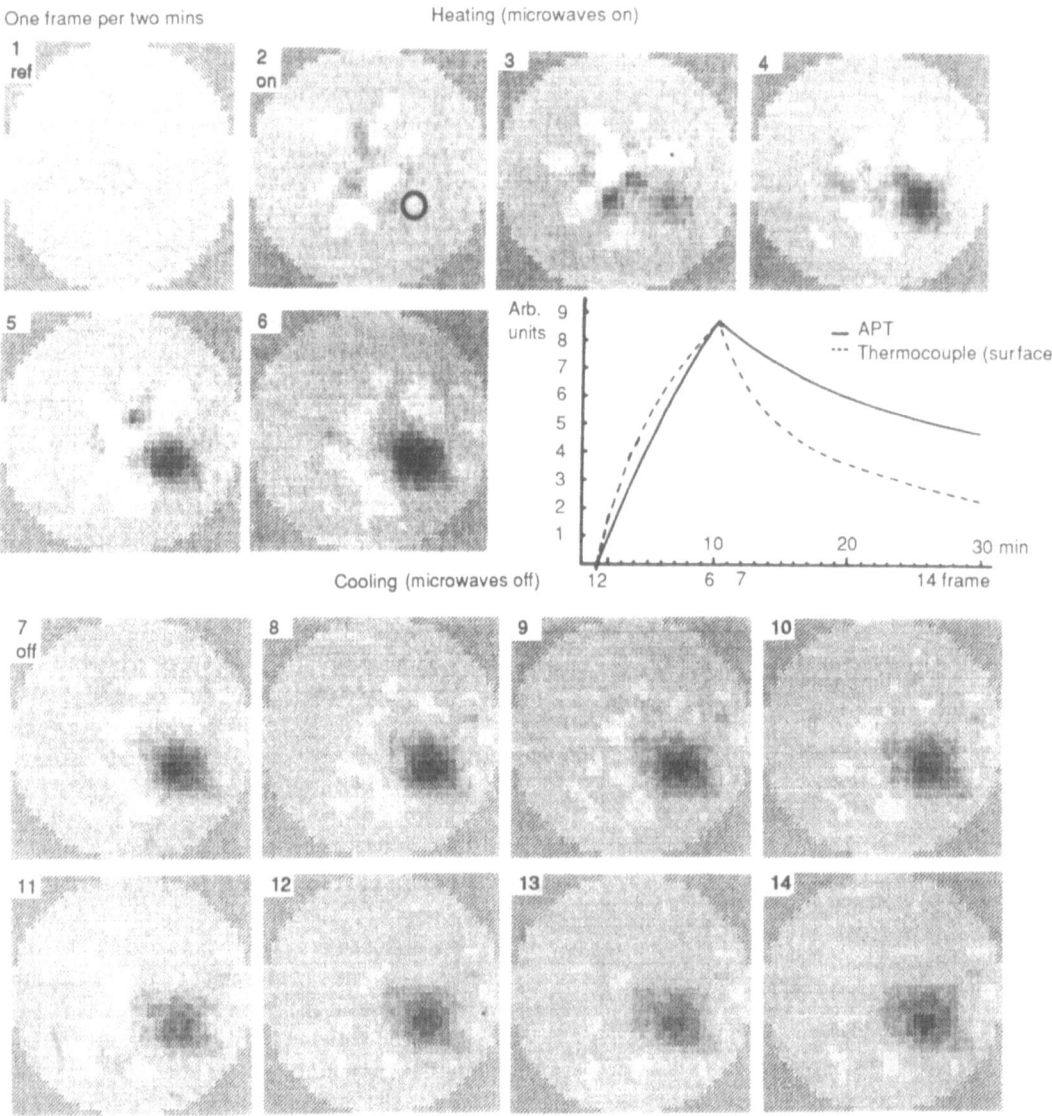

Fig. 2.38. Electrical impedance images of thermal changes in an agar phantom induced by microwave heating of the surface (○ marks the location of the applicator; increasing tempera-ture = white-to-black). A 9°C rise in the temperature was obtained in 10 min, followed by 30 min cooling

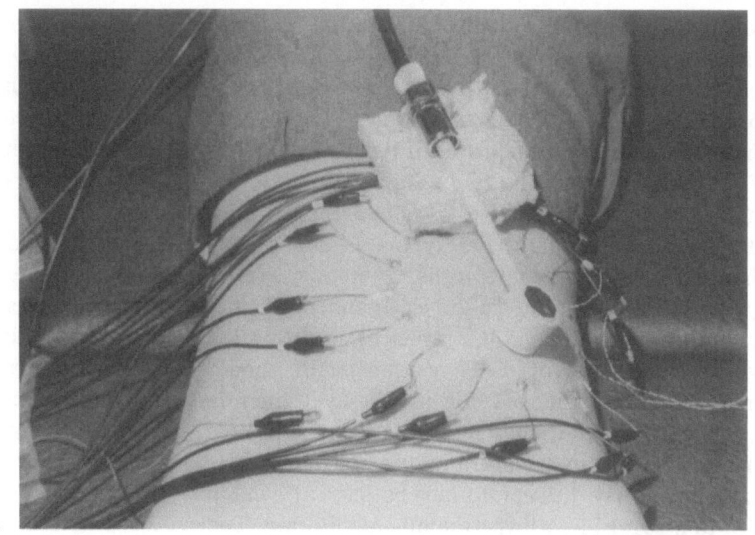

a

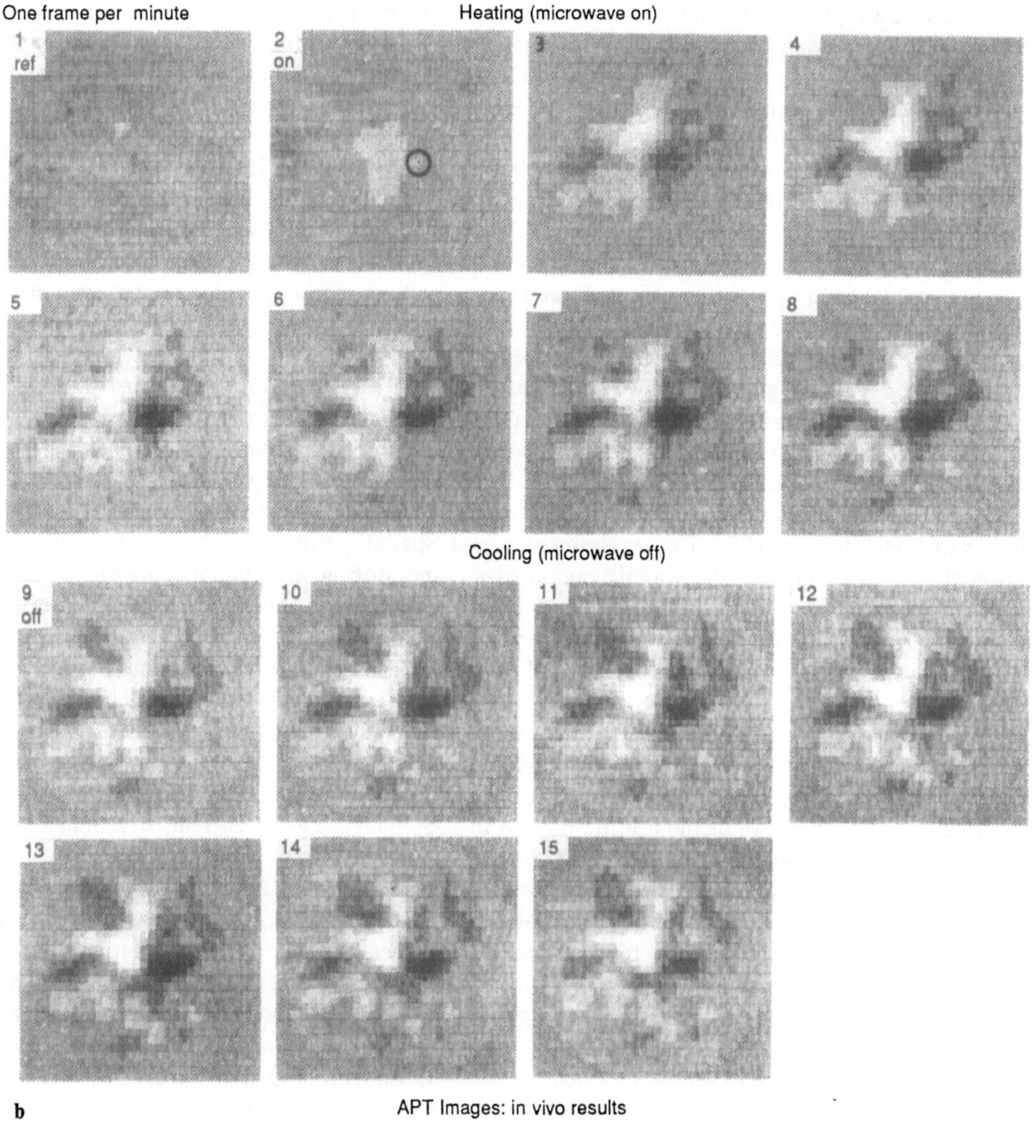

One frame per minute
Heating (microwave on)
1 ref
2 on
3
4
5
6
7
8
Cooling (microwave off)
9 off
10
11
12
13
14
15
APT Images: in vivo results

b

Fig. 2.39. a Experimental arrangement for the in vivo thermal imaging of a heated region in the scapula region of a volunteer.
b Impedance images for the in vivo experiment. A 7 °C rise in temperature was obtained in 8 min, followed by 9 min cooling. (By courtesy of J. Conway)

potential profiles by extrapolating from changes which take place during the respiratory cycle, where the resistivity of lung tissue can change from $10\,\Omega$-m to $20\,\Omega$-m. This 100% change in resistivity yields changes in the peripheral profile of typically only 5%. If we require a temperature resolution of, say, $1\,°C$, the system must be sensitive to changes of the order of 2% in the total lung resistivity; that is, about 0.1% change in the peripheral profile. Therefore, typical system sensitivities should allow a central temperature resolution of around $1.5\,°C$. It can be seen that the typical sensitivity of EIT measurements should preferably be augmented for use in hyperthermia monitoring, which usually means an increase in time of measurement to average a number of data sets.

Experiments have been performed both in vitro and in vivo to assess EIT for hyperthermia monitoring. Two different geometries have been employed. In the first, the electrodes are placed on a plane surface in order to map heating by applicators placed on the surface within the electrode ring [30]. Figure 2.38 shows images obtained when heating the surface of an agar phantom using a small microwave diathermy applicator positioned off-center and fed by 15 W at 2.45 GHz. Sixteen electrodes are placed in a circle of diameter 15 cm. The figure compares the image values averaged over the heated region with a thermocouple measurement taken just below the agar surface, under the applicator. It can be seen that the image values represent the heating and cooling of the region below the applicator. The disagreement between the curves may be due to the fact that the imaging currents flow in a given thickness of the agar and thus the image is an average over a certain depth, whereas the thermocouple gives a surface measurement which must heat and cool more quickly. This experiment was repeated in vivo by heating the scapula region of a volunteer (Fig. 2.39a). The images (Fig. 2.39b) show a greater degree of image artifacts which may be due to a combination of body movement and blood flow changes.

In a second geometry, one more closely analogous to X-ray tomography, electrodes are placed around the circumference of a cylindrical structure [31, 48]. Figure 2.40 shows an image of the heating pattern in agar of a small applicator operating at 2.45 GHz which is heating between two electrodes. Figure 2.41 shows an image of the capacitive heating at 13.56 MHz of a agar cylinder of 16 cm diameter. Saline boluses couple power into the agar, each consisting of two separate halves placed above and below the measurement electrodes. Contact areas were $11\,cm \times 7\,cm$ and $15\,cm \times 15\,cm$ to produce greater heating near the smaller electrode. This latter experi-

ment was reproduced in vivo by heating the thigh of a volunteer with a 175-W signal (Fig. 2.42). The image shows greater heating near the smaller electrode, as would be expected. Image artifacts are again more evident than in phantom experiments.

The imaging systems employed in all these experiments were very similar: 16-electrode systems with current drive of 5 mA at 50 kHz. Data acquisition times per frame were 8 s and 10 s (averaging 100 measurement cycles) with image display available within 30 s. The spatial resolution was approximately 10% of the circular array diameter, of the order of $1-1.5\,cm$, and the temperature resolution was around $1\,°C$ in vivo and around $0.2\,°C$ in phantom experiments.

2.6.4.5 Discussion

The temperature resolution of available measurement systems, from $0.2°$ to $1\,°C$, should be good enough for this technique to be used to control hyperthermia treatments. Temperature resolution could be improved by increasing the number of data sets collected or lengthening the integration time. In both cases, the data acquisition time would be lengthened.

The present data acquisition time could be reduced by a factor of 10 by changing from serial to parallel voltage measurements, at the cost of only slightly greater system complexity. However, slightly extended data acquisition times are less of a problem for EIT than for some other noninvasive measurement techniques, as it has not so far been found necessary to switch off the heating in order to collect data. Nevertheless, it is likely that much faster EIT systems will be available in the near future.

Spatial resolution is adequate for hyperthermia monitoring, although it is possible to further improve this factor by increasing the number of electrodes. For example 128 electrodes could give a spatial resolution of 2.25 mm in a 15 cm diameter area. Such large increases in electrode numbers could prove impractical, but a system with 32 electrodes is being considered, which would also yield an increase in sensitivity. Some improvement in sensitivity could also be gained by increasing the current drive magnitude.

It is clear from the results presented in the previous section that EIT images suffer from increased noise levels when measuring in vivo. Since the electrodes are placed in contact with the patient, movement due to respiration, etc. can cause artifacts to the same order as those measured for temperature changes. It is not clear at present how such artifacts can be avoided. Employing a greater number of measurement cycles to average out these variations may be a solution,

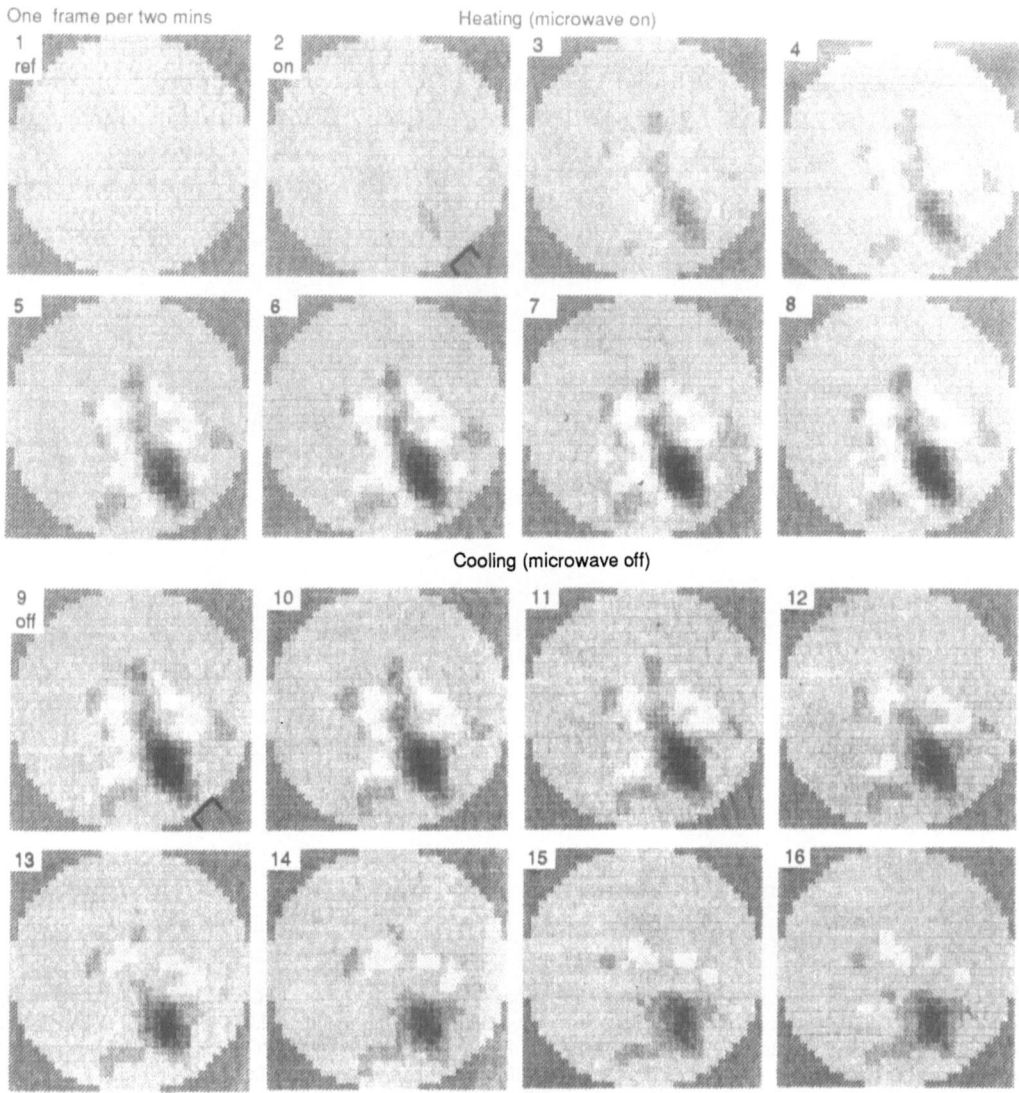

Fig. 2.40. Impedance images of the heating of a cylindrical agar block with a small applicator at 2.45 GHz placed between two electrodes tangential to the cylinder surface. (By courtesy of J. Conway)

although at the cost of longer measurement times. Alternatively, if much faster data acquisition were possible, as suggested above, movement artifacts could be reduced. The images are also sensitive to variations in blood flow. Imaging of a forearm on occlusion of the venous return by a tourniquet indicated at 1% – 2% decrease in resistivity, equivalent to a 1 °C change in temperature. Few data are available at present on the variation of blood flow with temperature, and more investigation is needed in this area before any conclusions can be drawn.

An effect, which has yet to be explained, has been noted when microwave heating of phantoms or patients is imaged by EIT [31]. In some cases, the image

fails to return to a uniform distribution as the temperature of the heated volume returns to the reference level on cooling. This effect has not been noted in conjunction with any other type of heating, and investigation continues into this phenomenon.

The slice thickness encountered in EIT is fairly large, of the order of one-third of the electrode ring diameter. This leads to a low spatial resolution in the plane perpendicular to the electrode ring. Some improvement could be achieved by employing two or more parallel rings. Many reconstruction algorithms are based on current flow paths calculated in homogeneous media. The presence of strongly inhomogeneous structures such as bones distorts cur-

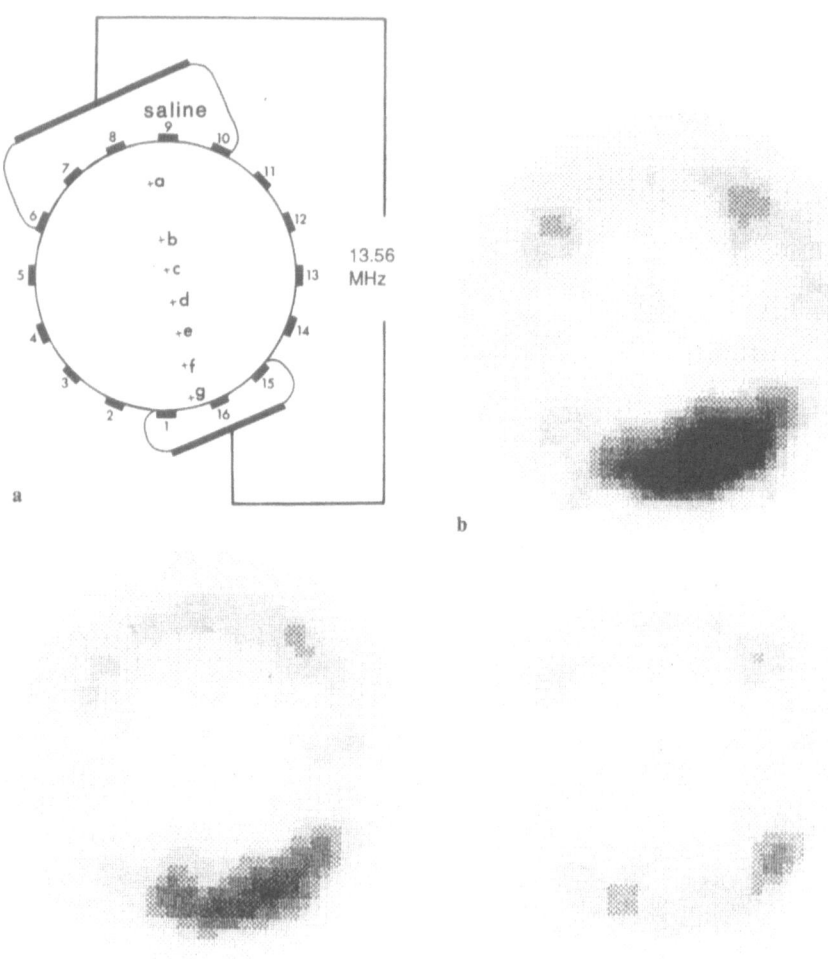

Fig. 2.41. a Schematic diagram of radiofrequency capacitive heating of an agar cylinder showing position of heating electrodes, saline boluses, imaging electrodes (positions 1–16), and thermocouples (positions a–g). **b** Impedance images following 6 min heating, showing greatest heating near the smaller electrode. The thermocouples recorded temperature rises: a: 1.5 °C; b: 1.7 °C; c: 2.0 °C; d: 2.5 °C; e: 2.6 °C; f: 5.9 °C; g: 8.5 °C. **c** After 9 min: a: 2.0 °C; b: 1.7 °C; c: 1.9 °C; d: 2.7 °C; e: 3.4 °C; f: 6.0 °C; g: 3.2 °C. **d** After 16 min: a: 2.1 °C; b: 1.6 °C; c: 1.9 °C; d: 2.4 °C; e: 3.8 °C; f: 5.4 °C; g: 3.2 °C. (By courtesy of H. Griffiths)

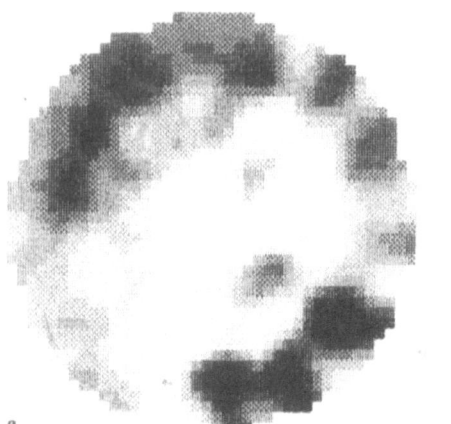

Fig. 2.42a, b. Impedance images following heating of a thigh. Imaging and heating electrodes positioned as in Fig. 2.41a, except in this case the larger saline bolus extended from electrodes 7 to 11. **a** Image recorded within 40 s of cessation of RF power. **b** Image after a further 8 min. (By courtesy of H. Griffiths)

rent paths, causing inaccuracies in the reconstructions. Some progress is being made towards taking these effects into account [3].

Compatibility of the measurement system with hyperthermia systems is easily achieved. The measurement accuracy does not appear to be degraded by the presence of strong heating fields. Four different combinations of measurement system and heating system from 525 kHz to 2.45 GHz have already been demonstrated and more can be imagined. The large slice thickness found with EIT may in some cases be employed to advantage if the measurement electrodes cannot be placed in the same plane as the heating applicators.

Perhaps the greatest advantage of EIT compared to other modalities it its relatively low cost and simplicity of operation, using standard electrode types (e.g., ECG electrodes). Even if the large research effort being pursued in this area does not eventually produce a technique which is sensitive enough to control hyperthermia on its own, its use for imaging of heating patterns, detecting and pinpointing unwanted hot-spots, etc. would constitute an important additional control factor for hyperthermia well within the financial means of most hyperthermia clinics.

2.6.5 Microwave Imaging

2.6.5.1 Historical Aspects

Contrary to other tomographic modalities such as X-ray tomodensitometry and NMR imaging which are able to take profit of the experience of many years in classical radiography or of laboratory experiments, microwave imaging has started quite recently. Some possible explanations are probably an a priori idea of image quality and more definitely the lack of appropriate recording devices for microwave field distributions. The credit for having produced the first really convincing images demonstrating the possible usefulness of microwave imaging for biomedical purposes belongs to Drs. Larsen and Jacobi [66, 67, 74] from the Walter Reed Army Institute. The key point of their approach was to use the immersion technique consisting of immersing the biological target in a medium, thus minimizing the reflection losses and also improving the penetration, reducing the parasitic effects of external paths around the target, and contracting the wavelength according to the index of the immersion medium [66]. The high dielectric constant of high water content media provides a significant improvement in spatial resolution, circumventing preconceived ideas on this aspect.

The first images were obtained by transmission, moving simultaneously the emitting and the receiving antennas. The basic idea was, after repeating the previous experiment for various orientations of the target, to reproduce in the microwave region the situation of X-ray tomography. Unfortunately, diffraction effects occurring during the propagation through inhomogeneous media and the impossibility of realizing well-collimated beams rendered useless the existing algorithms for X-rays and, to some extent, for ultrasonic waves. Time of flight procedures have been considered in order to isolate the most direct path, but the results have shown that the image quality is not significantly improved [67].

Such a situation led to the consideration of more complicated tomography algorithms for taking into account diffraction phenomena. As explained later, the algorithms present some serious weaknesses which render more difficult the interpretation of microwave images, except in some particular situations. A lot of effort has been devoted to the development and the improvement of such algorithms without reaching a really satisfactory situation [1, 35, 41, 69, 82, 87, 96]. The considered algorithms have been adapted to three main experimental configurations (Fig. 2.43). The first, called *planar* in the following, consists in illuminating the target by means of a plane wave, as uniform as possible, and collecting the scattered field over a plane surface normal to its propagation direction. Such measurements repeated for several orientations provide the required set of data for tomographic reconstruction. The second configuration is called *cylindrical*. In this case, the experimental arrangement consists of a circular, or cylindrical, array of small antennas which are successively activated as emitting antennas while the others are used as receiving antennas. Many other configurations could be used. The fact that it minimizes the number of necessary antennas makes the so-called *crossed-array configuration* attractive; this consists of two orthogonal linear arrays, the elements of which are, as previously, used in the emitting or receiving mode.

2.6.5.2 Equipment Considerations

The lack of convenient means of recording explains to a large extent the late development of microwave imaging. In their first experiments in 1979 [74], Drs. Larsen and Jacobi needed a few hours just to record the field transmitted by an isolated organ such as a horse kidney. The limitations of mechanical scanning are evident. Classical arrays of antennas were devel-

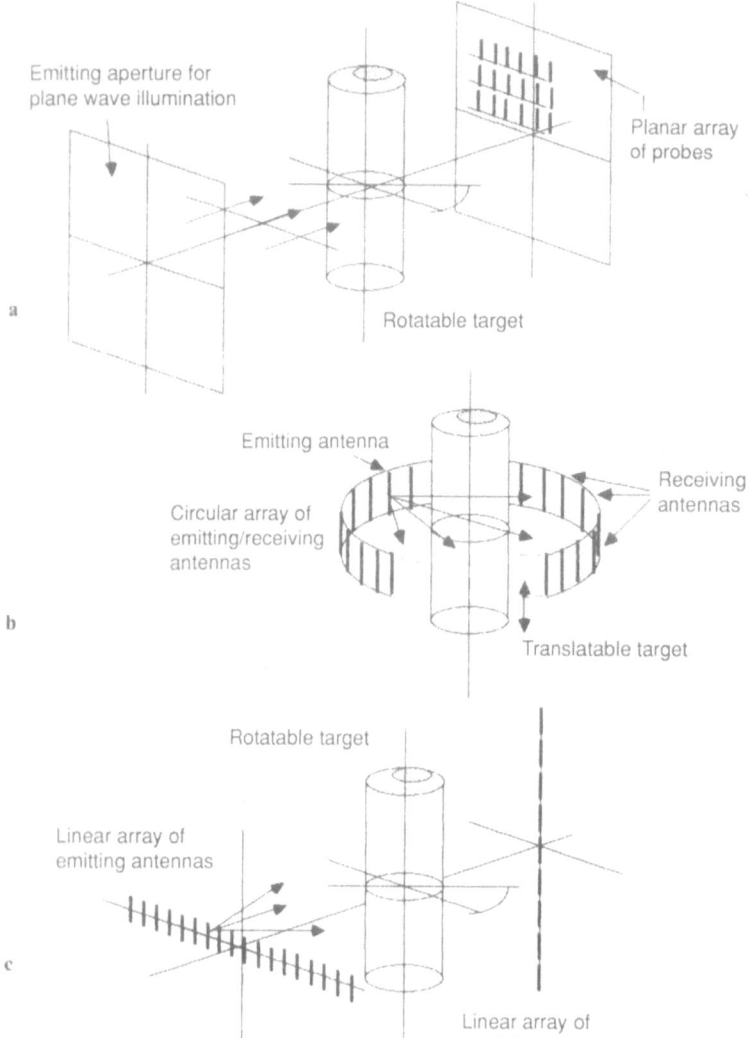

Fig. 2.43. Three main experimental arrangements for microwave tomography: **a** planar array; **b** circular array; **c** crossed linear arrays

oped later [51]. "Classical" means here that each antenna is connected to the same receiver through a microwave multiplexer. The need for a microwave link for each probe renders difficult the achievement of the spatial sampling interval required by further tomographic reconstruction. For diffraction tomography, a convenient sampling spacing is approximately one-half of a wavelength. In classical arrays developed by Drs. Larsen and Jacobi, the individual probes consisted of ridged rectangular waveguides filled with water. Each waveguide was equipped with a coaxial to waveguide transition. The waveguide size only allowed one wavelength spacing, resulting in so-called grating lobes in antenna language [51]. The existence of such grating lobes means that the array cannot be sensitive to only one point but to a finite number of points variously located inside the target.

The modulated scattering technique (MST) has been used to achieve a good compromise between rapidity, sensitivity and cost. According to this technique, the retina consists of an array of small dipoles loaded by a nonlinear device such as a PIN diode [e.g. 13]. Modulating one diode at a low frequency rate produces a modulated perturbation which can be collected by means of an especially designed antenna, called a collector. The microwave signal at the collector output is modulated and directly proportional to the field at the corresponding dipole location. By means of a sequential addressing of the modulation signal to the different diodes, it is possible to rapidly record the field over the retina without a microwave multiplexer. By comparison with classical arrays, MST arrays allow the realization of smaller elementary antennas in order to accommodate the sampling rate required

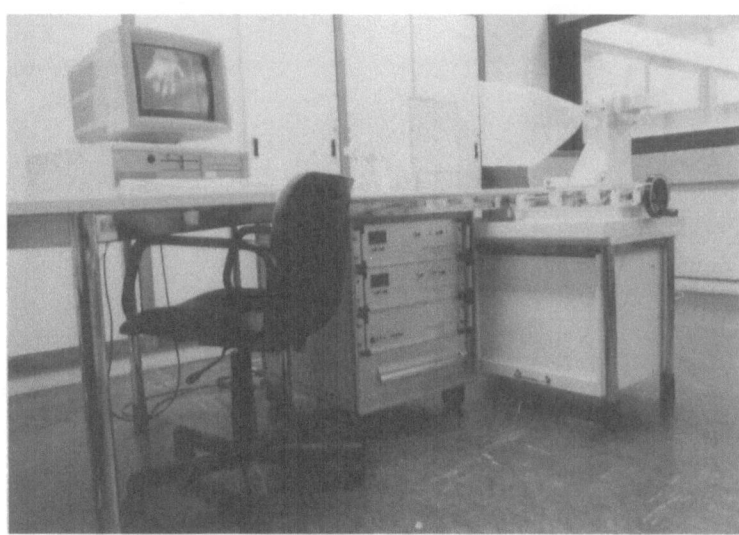

Fig. 2.44. Planar microwave equipment for biomedical applications at 2.45 GHz. (By courtesy of Satimo)

by further numerical processing. Figure 2.44 shows a microwave planar camera realized by using MST. The experimental equipment initially designed to operate in water according to the immersion technique can also be used for phantom experiments.

Figure 2.45 shows a circular array which uses the modulated multiplexing technique (MMT) [68]. Such a technique can be considered as located midway between classical and MST techniques. The experimental procedure consists in activating one antenna as an emitting antenna, while the others are used as receiving antennas. The source is connected to the selected emitting antenna and the receiving antennas to the receiver via a microwave multiplexer. The performance of the multiplexer can be significantly improved by periodically switching the selected channels at two different frequencies, for the source and the receiver, respectively. The measurement signal can then be isolated and discriminated against non-modulated signals by means of a coherent detection at the beat frequency corresponding to the difference between the two previous switching frequencies.

In all cases, these experimental arrangements suffer from the need for target immersion. The use of rigid water tanks imposes serious limitations on the handling of various targets. On the other hand, boluses pose serious contention difficulties and their practical usefulness has not yet been fully investigated.

Fig. 2.45. Circular scanner for microwave tomography at 2.45 GHz (Jofre et al. [68])

2.6.5.3 Spatial Resolution Capabilities

Until now, microwave imaging systems can be considered as diffraction limited. To some extent, an optical analogy allows the evaluation of their spatial resolution capabilities. Figure 2.46 shows a typical transmission arrangement in which the observed object is illuminated by a parallel and uniform beam. The image of the object – which is implicitly assumed to be transparent at the considered wave-

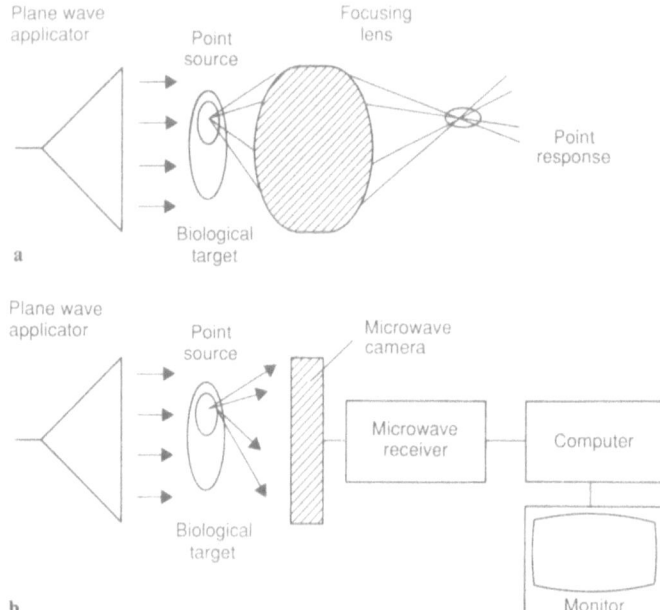

Fig. 2.46. a Optical analogy of microwave imaging and **b** the practical realization

length – is obtained in the focal plane of a focusing system. Even in the optical case of microscopes or telescopes, it is well known that the image of one object point is not a point but a small spot. Two different points will then be resolved if their respective spots do not overlap too much. More quantitatively, the Rayleigh criterion predicts, according to diffraction theory, that the distance between these image points must be greater than one-half of a wavelength in the observed medium. This distance is sometimes called the transverse resolution. On the other hand, due to the finite field depth of the focusing system, the image is not entirely contained in its focal plane but extends in a direction normal to this plane. Applying the same overlapping criteria for two different object points, it is possible to show that their distance along the axis must be greater than, roughly, two to three wavelengths, partly depending on the distance to the recording area [87]. The longitudinal resolution, which is nothing other than the field depth, is then worse then the transverse resolution.

Changing the focusing plane allows, to some extent, the achievement of tomographic reconstructions. As a matter of fact, the image in a given plane corresponds to an integrated slice of the object. It can be seen that spatial resolution can be rendered uniform by using images corresponding to different orientations of the object.

With respect to this optical situation, the microwave case exhibits aggravated diffraction phenomena because the dimensions of the object – or those of some details – as well as the dimensions of the focus-

ing system can no longer be considered as very large with respect to the wavelength. Furthermore, the object is generally more lossy. These two factors contribute to the degradation of the spatial resolution. For the sake of flexibility, the focusing system consists of a retina, which allows the recording of the amplitude and the phase of the wave scattered by the object, and of a numerical processor to simultate the effect of the focusing system over the scattered wave front. As a result the ultimate performance is dependent on the signal to noise ratio during measurements on the retina, and on the stability of the reconstruction algorithms.

2.6.5.4 Specific Difficulties in Diffraction Tomography

Specific difficulties result from the nonlinear dependence of the scattered field with respect to the ideally desired complex permittivity. As previously indicated, the linearity only exists between the scattered field and the equivalent currents which introduce the local total field dependence of the reconstructed images. Whatever the experimental arrangement – planar, cylindrical, or crossed arrays – the quantity to be reconstructed is the normalized equivalent current distribution which, assuming nondepolarization, is given by:

$$\mathbf{J}_{nor}(\mathbf{x}) = j\omega\{\varepsilon(\mathbf{x}) - \varepsilon_a\}\mathbf{E}_t(\mathbf{x})/\mathbf{E}_I(\mathbf{x}) \ . \tag{2.58}$$

Knowing that the linear inversion problem is much easier to solve than the nonlinear one, the first question is whether the normalized equivalent current distribution provides an adequate representation of the observed media. "Adequate" means that the shape and localization of the different anatomical structures is reproduced, that homogeneous parts produce corresponding homogeneous areas in the image, and that image intensity can be related to some parameter of clinical interest.

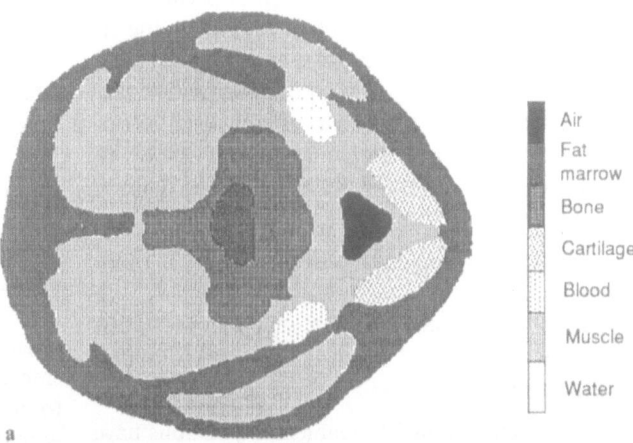

Fig. 2.47 a – c. Two-dimensional numerical simulation of the equivalent current distribution in a neck cross-section in multiview experiments. **a** Neck cross-section from CT X-ray scan; associated meshing for numerical processing, trough moment method, a eight illumination angles. **b** f = 1.225 GHz, right: scalar (*right*) and vector (*left*) addition of partial views. **c** f = 2.45 GHz, right: scalar (*right*) and vector (*left*) addition of partial views. (de Talhouët [34])

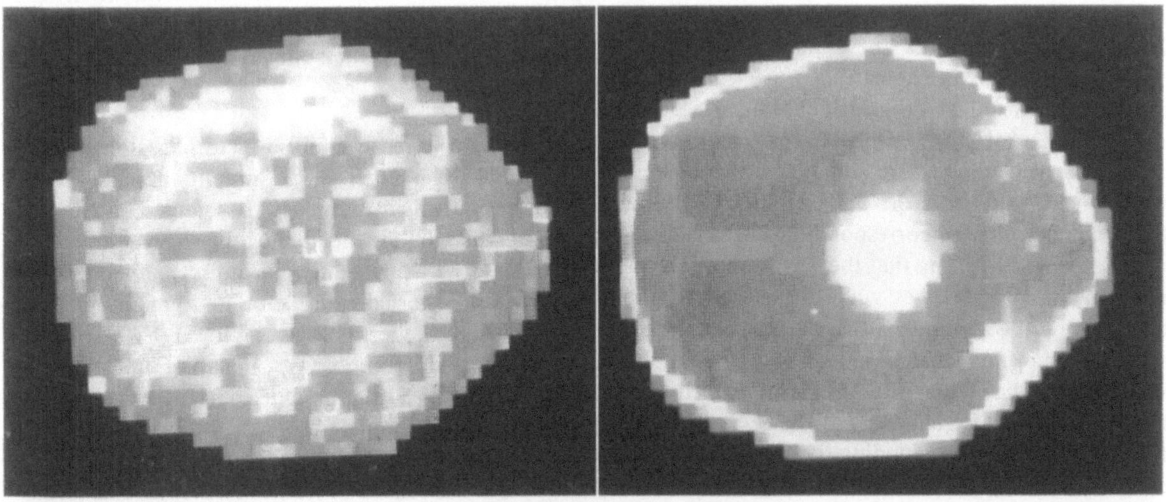

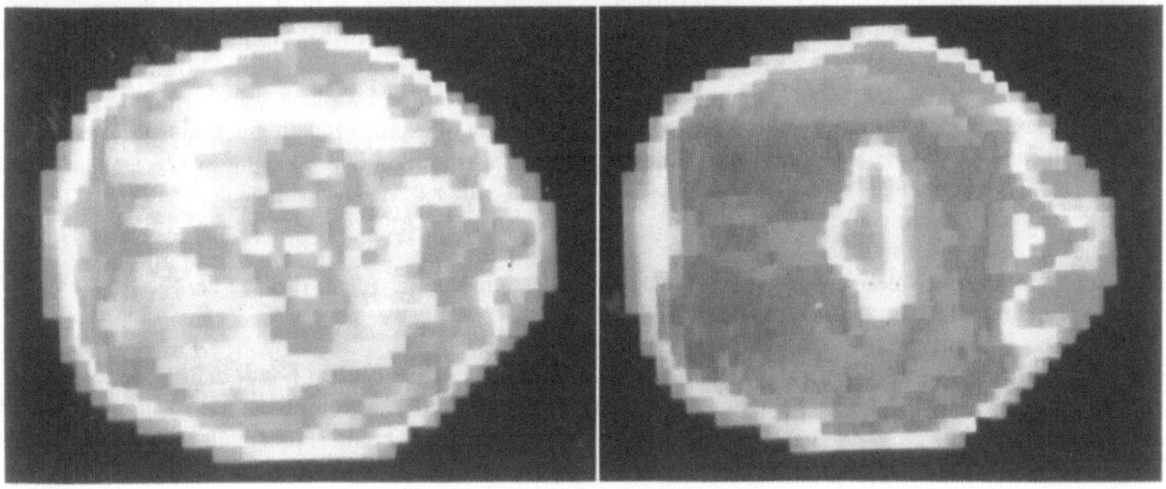

The second question of importance is whether it is possible to determine accurately the equivalent current distribution by means of suitable processing of scattered field data. This question is not so simple to answer. As previously shown, true tomographic reconstructions require multiview experiments. Although, for evident reasons, the complex permittivity is not affected by the angle of incidence of the illuminated wave, the local incident and the total field depend on it. As a result in the general case, the different data obtained at different orientations do not correspond to the same normalized current distribution! However, two favorable situations are as follows:

1. The observed target verifies to so-called Born's approximation or, roughly speaking, is quasi-homogeneous.
2. One is only interested in differential imaging to display the variations of a slowly and weakly changing parameter.

The two aforementioned fundamental questions have been answered relatively positively, mainly by means of numerical modeling of complex structures. Figure 2.47, for instance, gives the simulated normalized equivalent currents in the case of a neck region obtained by means of two-dimensional numerical modeling, at 1.225 GHz and 2.45 GHz for eight different views [34]. The images were obtained in two different ways consisting in scalar or vector addition of the partial amplitude images corresponding to the different views. It is worth noting that the vector addition does not provide good images while the scalar addition allows good recognition of the different anatomical structures, both at 1.225 GHz and at 2.45 GHz. It is believed that the high contrast between dielectric permittivity of the different parts of the neck, as well as the losses of the corresponding tissues, explains such satisfactory images from which it can be seen that the local total field dependence of the equivalent currents does not perturb too much the complex permittivity content. This result is quite important because it demonstrates that even qualitative imaging, based on equivalent current reconstruction, is able to provide useful information. Such a demonstration should stimulate the development of imaging equipment capable of reconstructing as well as possible the equivalent current distributions.

It is not yet completely clear how the microwave images can be combined in order to get the most significant results. Amplitude and phase images can be reconstructed from the measured wavefronts. Furthermore, images can be obtained in the nominal polarization or in the cross-polarization. As in other electromagnetic areas, polarization information is believed to provide a means of discriminating significant details from artifacts or of improving the global contrast. However, as yet the introduction of the polarization aspects in the reconstruction algorithms cannot be considered completely satisfactory. A lot of work still has to be done in this area.

2.6.5.5 Optimization of the Operating Frequency

Improving the spatial resolution would imply operating at the highest possible frequency. In fact, however, the frequency elevation is limited by considerations relating to sensitivity. Indeed, the losses in high water content tissues rapidly increase with frequency, resulting in a drastic attenuation of the transmitted signal. In order to maintain a given signal to noise ratio, one may increase the source power — with evident limitations to avoid thermal effects — or reduce the noise level. In microwave tomography the main source of noise is the thermal noise at the receiver input. The noise power is given by:

$$N = F k_B T \Delta f \tag{2.59}$$

where F is the receiver noise factor, k_B Boltzmann's constant, T the target temperature in °K, and Δf the receiver equivalent frequency passband. This thermal noise is then mainly dictated by the receiver noise factor F — or, equivalently, its noise temperature — which can be varied by a factor $1-10$ approximately, according to the technology employed. Even with the most desirable technology, the noise level can be reduced by increasing the time constant, at the cost of a reduction of the measurement rate.

The signal power can be simply estimated by considering a homogeneous target consisting of a homogeneous water layer of thickness L. If the applicator aperture A is much larger than the wavelength λ in water, the microwave propagation in water is approximately governed by plane wave properties, mainly consisting of exponential attenuation. More accurately, this assumption is valid as far as L is smaller than the so-called Rayleigh distance D_R such that:

$$D_R = L^2/(2\lambda) \ . \tag{2.60}$$

The signal power S can then be derived as:

$$S = P_{inc} \sigma_p \exp(-2\alpha L) \tag{2.61}$$

where P_{inc} is the incident power density and σ_p is the absorption cross-section of the measuring probe at a given sampling point. For small probes — as are generally used in order to accommodate a sampling

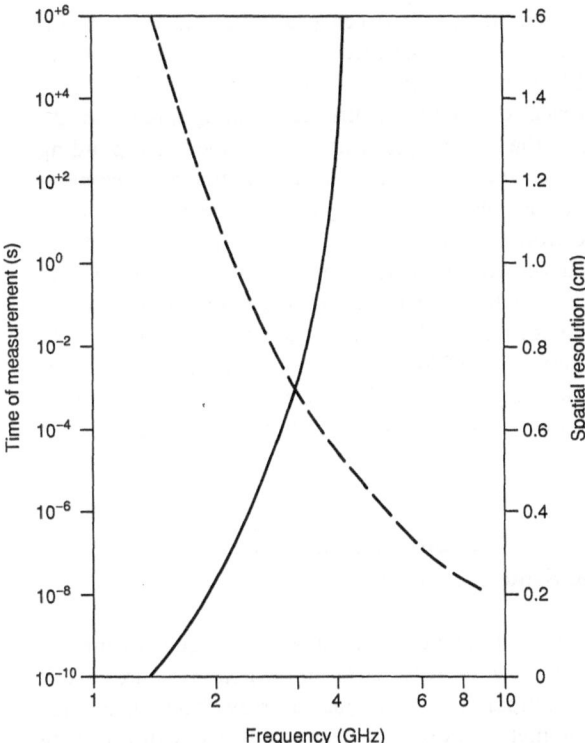

Fig. 2.48. Time constant τ and spatial resolution Δx in water versus frequency. $T = 300\,K$, $F = 1.3$, $P_{inc} = 10\,mW/cm^2$, $L = 20\,cm$, S/N = 50 dB

space of one-half of a wavelength – a good approximation of σ_p is $\lambda^2/(4\pi)$.

The noise equivalent bandwidth of the receiver is approximately equal to the inverse of its time contant τ, which can be considered as the minimum time required to make a measurement.

This time constant τ is given by [14]:

$$\tau = \frac{8\pi k_B F T}{P_{inc}\lambda^2}(S/N)\exp(2\alpha L) \ . \qquad (2.62)$$

Figure 2.48 gives a typical example illustrating the corresponding variations of τ and of the spatial resolution as a function of frequency for a given signal to noise ratio. This figure shows the drastic increase of τ with frequency.

Now, the choice of the operating frequency has to be determined according to the required performance in the configuration under consideration. A simple rule is to operate at the frequency minimizing the loss-angle tangent, which, as previously indicated, is the ratio of ε'' over ε': ε'' is directly proportional to the attenuation constant, and the spatial resolution is proportional to the wavelength in the vacuum divided by $\sqrt{\varepsilon'}$. Minimizing the loss-angle tangent tgδ maximizes the spatial resolution to attenuation constant ratio.

Despite its simplicity, this rule has been validated by more sophisticated considerations. One of the possibilities was indeed to operate at lower frequencies for which the attenuation constant in living tissues is much lower and which allow oversampling with respect to usual sampling theorem. Such oversampling provides, at least in principle, access to evanescent waves diffracted by the target and was expected to yield much better spatial resolution than the diffraction limit of one-half of a wavelength. This approach, which is relevant to so-called superresolution problems, does not prove very efficient if the target is located at distances greater than a few centimeters. But it is worth noting that a 0.3 cm resolution could be attained at up to 1 cm depth at 500 MHz, for which the wavelength is, however, 60 cm [34].

2.6.5.6 Some Results

As already indicated, microwave imaging is a new technique as compared with other more classical modalities. The results which are presented below merely have the purpose of illustrating its most promising features. These results have been procured from research conducted with mechanical scanners or with arrays of probes. Within the framework of a special evaluation procedure conducted in France (GBM/TEP Procedure), two industrial prototypes, operating at 2.45 GHz in accordance with the planar geometry, have been used in order to confirm theoretical predictions and to define the limits of usefulness of these prototypes in various configurations. In Spain, a prototype circular array has recently been produced. For evident reasons, results are more abundant from groups equipped with probe arrays, with which measurement durations are of the order of a few seconds, than from groups with mechanical scanners, which have a measurement duration of a few hours.

Spatial resolution capabilities in homogeneous media, in both the transverse and the longitudinal plane and in single view as well as multiview protocols, are clearly illustrated by Fig. 2.49. For single view measurement, the experimental values of $\lambda/2$ and 2λ correspond approximately to the theoretical values predicted on the basis of diffraction limitations.

The microwave cameras have been combined with several heating modalities to demonstrate their thermal sensing capabilities as well as their compatibility with heating equipment. More particularly, the case of capacitive heating has been considered as representative of deep or semideep hyperthermia sessions. To

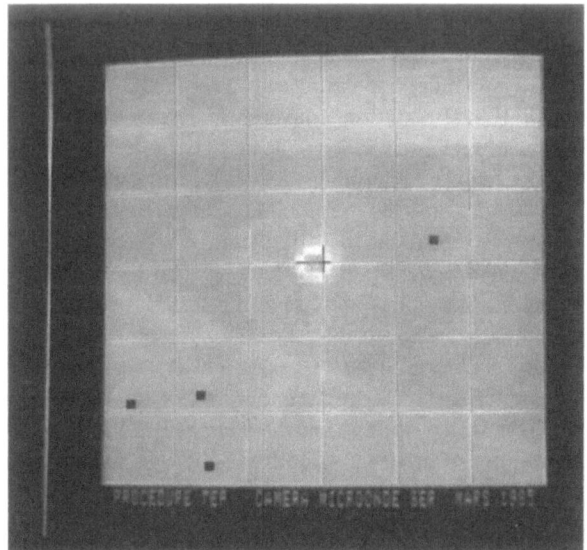

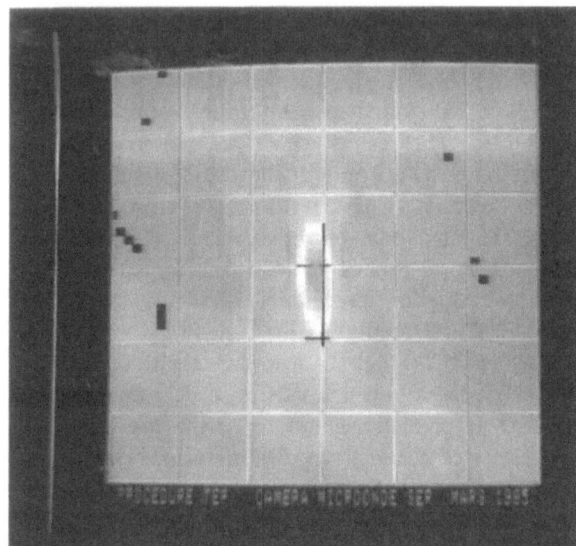

Fig. 2.49a, b. Transverse and longitudinal resolution in single view experiment obtained at 2.45 GHz with the planar microwave camera shown in Fig. 2.44, scale 3 cm/div

this purpose, a $21 \times 30 \times 16$ cm^3 muscle equivalent gel has been heated by means of 13.56-MHz standard equipment. The complex dielectric constant of the gel is $\varepsilon' = 65$, and $\varepsilon'' = 14.5$ at 28.5 °C. The spacing between two 10×10 cm^2 electrodes, located at two opposite faces of the phantom, is 21 cm. Microwave imaging is obtained at 2.45 GHz by transmission through cross-illumination over the 16 cm thickness of the gel (Fig. 2.50). Non-perturbing thermometry is achieved during heating by means of a Luxtron fiberoptics probe at some points located in the median plane.

The relationship between peak microwave image intensity and temperature is linear, as shown in Fig. 2.51, with an accuracy of 0.4 °C. Figure 2.52 gives sequences of successive differential images reconstructed in the median plane and in a plane corresponding to the edge of the electrodes, at different temperatures. Preferential heating in the close vicinity of the electrodes is well visible and the image reproduces well-known temperature distribution patterns. Other experiments have been conducted by using external heat sources, such as water tube or serpentine at temperature T + ΔT in a water tank at temperature T, demonstrating the possible detection of temperature increases of the order of 1 °C through more than 10 cm gel thickness.

2.6.5.7 Combining Heating and Monitoring at the Same Frequency

Microwave imaging presents some unique advantages when it can be achieved from the heating source itself. A comparable idea has already been used for radiometry in order to heat and probe with the same antenna. The active imaging processes offer some additional flexibility. Indeed, it is possible to think of two successive sequences. Firstly, before any treatment, microwave imaging is able to indicate noninvasively the localization of the power density resulting from the applicator radiation. This possibility is particularly interesting in the case of multiapplicator systems, which pose the problem of optimizing the amplitude and phase feeding of each of the applicators. It is indeed very important to be sure that the applied radiation is focused, or at least concentrated, in the region of interest before applying the nominal heating power level. Secondly, after this power level is applied, it is important to follow the microwave image variations under temperature changes by means of differential imaging [29]. Figure 2.53 shows a prototype consisting of a semicircular array of eight dielectric filled waveguide applicators. Each applicator can be fed from an independent attenuator/phase shifter module. The array is located around a wet sand phantom with dielectric properties close to those of lung. A linear microwave camera is located on the phantom on the opposite side to the array. Figure 2.54 illustrates the possibility of visualizing the field distribution from measurement achieved by the camera. The focal point of the array is particularly well visible on the phase patterns. It is not surprising that the maximum of the field distribution is not located at the focal point but close to the waveguides' aperture due to the losses of the phan-

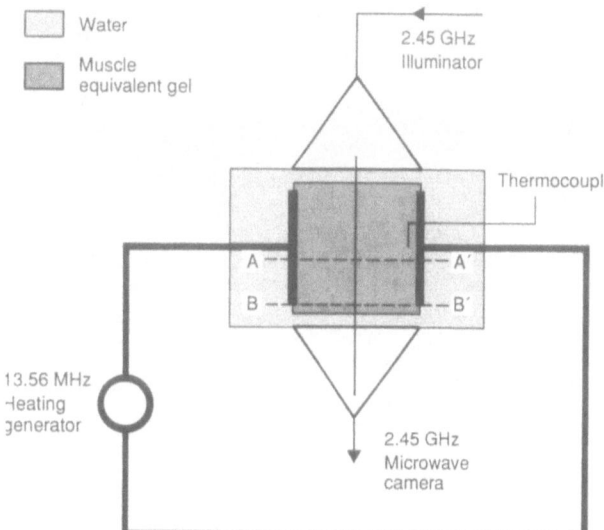

Fig. 2.50. Experimental setup combining capacitive heating at 13.56 MHz and transmission/single view microwave imaging at 2.45 GHz. AA′ and BB′ represent two reconstruction planes shown in Fig. 2.52

tom. It is worth noting that good agreement exists between measured and numerically predicted results. Figure 2.55 shows, by differential imaging, the heat propagation inside the phantom.

A more simple combination of heating and monitoring devices has been described more recently. It consists of two rectangular waveguides operating at 434 MHz [63]. During the heating cycle, both applicators are radiating (Fig. 2.56). More or less constructive interferences can be produced according to the tilt angle between applicators. During the control cycle, it is the transmission coefficient between the applicators which is measured. On the basis of interferences between directly transmitted waves and waves diffracted by tissue layers, it is shown that the modulus of the reflection coefficient is proportional to the temperature elevation measured at some points inside the target volume. In this setup, the measurement of the transmission coefficient is achieved by means of a network analyzer. Further studies are devoted to the localization of the temperature gradients between the two antennas.

Still more simply, a microwave camera appears an efficient tool for the design and testing of new applicators, provided their operating frequency is the same as the camera frequency. Assuming that the applicator to be investigated is radiating in a homogeneous medium, e.g., gel, the measurement of the radiated field by means of the microwave camera on the opposite side allows reconstruction of the field distribution at any intermediate point between the applicator and the camera planes (Fig. 2.57). Such a procedure avoids time-consuming mechanical probings of the field inside the phantom. The thickness must be large as compared to the penetration depth at the

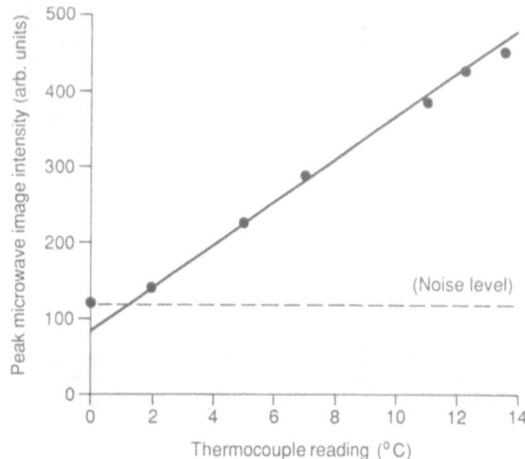

Fig. 2.51. Correlation between the peak intensity of the microwave image obtained in the plane AA′ and the measured temperature at the corresponding point (by courtesy of G. Gaboriaud, GBM/TEP procedure)

operating frequency in order to reduce parasitic multiple reflections between applicator and camera to an acceptable level. Figure 2.58 shows the field distribution of a commercial applicator measured with the microwave camera.

2.6.5.8 Comparison Between Active and Passive Microwave Imaging

The comparison between active and passive approaches is of prime importance in the sense that such a comparison relates a new technique to be developed to the only technique to be used in clinical situations.

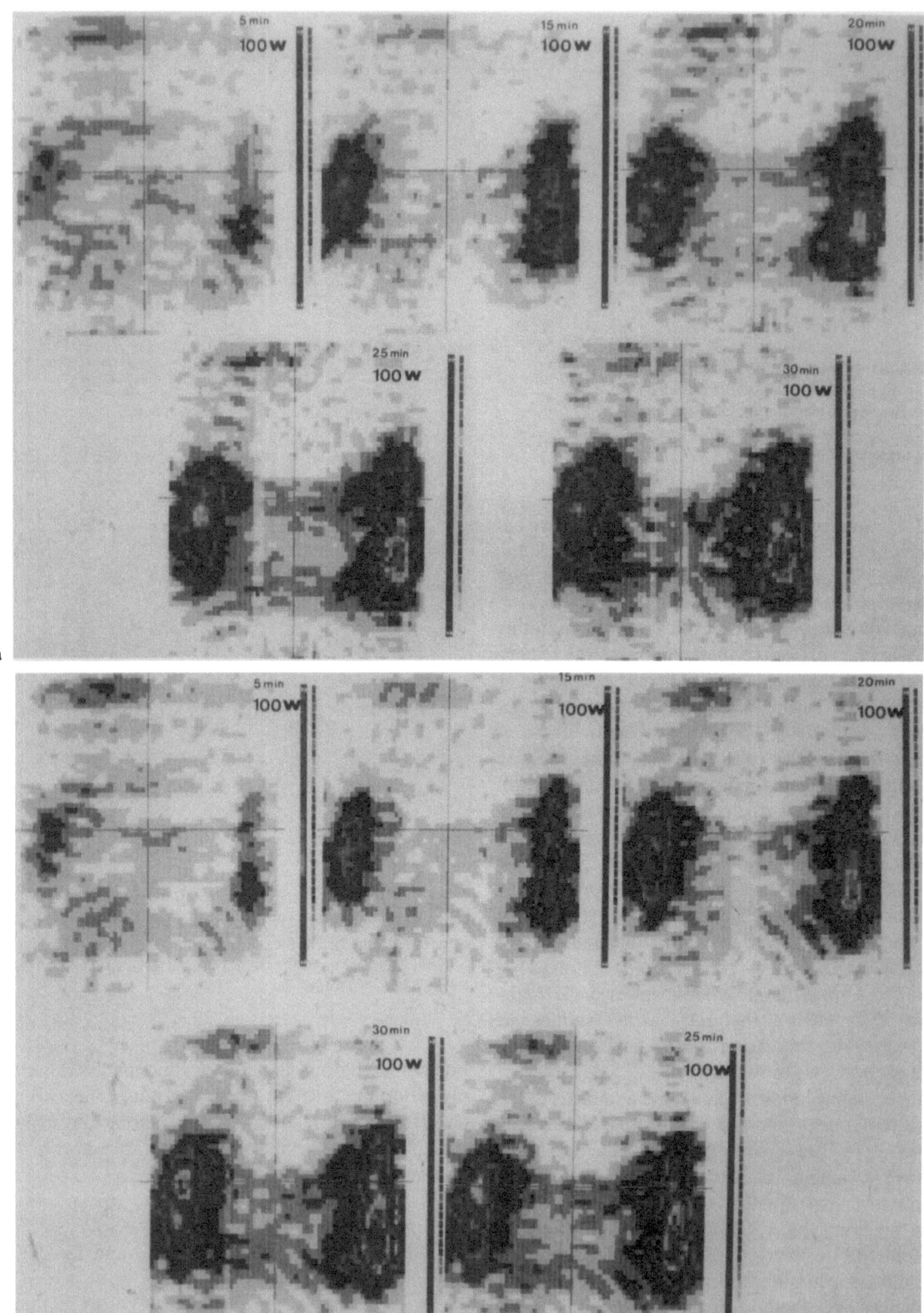

Fig. 2.52. Sequences of differential images obtained in planes AA′ (**a**) and BB′ (**b**) at different times, every 5 min after deliver-ing a power equal to 100 W by means of the RF generator (by courtesy of G. Gaboriaud, GBM/TEP procedure)

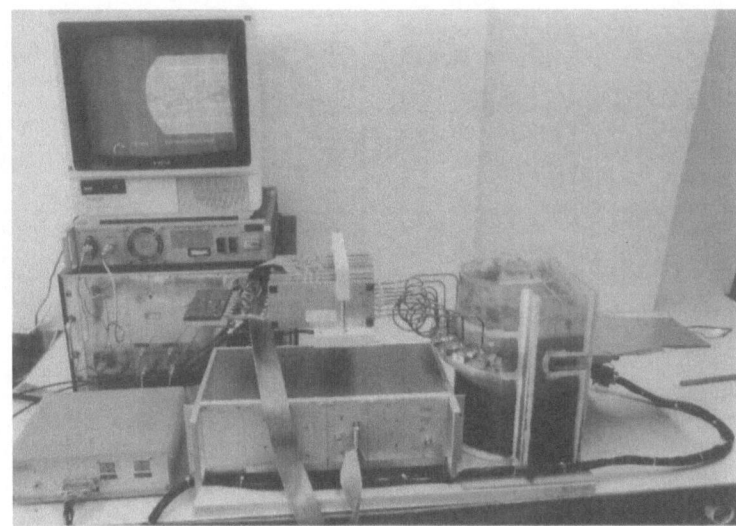

Fig. 2.53. Experimental setup combining microwave heating and imaging at 2.45 GHz. The applicator consists of a semicircular array of eight dielectric filled waveguides. The microwave camera is linear. (Coldefy [29])

If the merits of microwave radiometry are becoming recognized and its potentials better estimated, many questions are still open concerning active imaging. The first point to be elucidated concerns the popular feeling that radiometry provides measurement more closely related to temperature – and, hence, less sensitive to other parameters of a geometric or electrical nature – than does active imaging. This feeling is reinforced by the fact that radiometer outputs are calibrated in terms of temperatures directly converted in degrees Celsius. Such a situation has generated some misunderstanding as to the real capabilities of microwave radiometry in the context of noninvasive thermometry during hyperthermia treatments. It has taken some time for clinical users to realize that the same temperature increase at the radiometer output can be related to very different temperature profiles in the tissues under observation. The depth, the thickness, and more generally the shape and size of the thermal target introduce significant weighting, mainly through the emissivity in the relationship between the radiometer output and the temperature profile (see Sect. 2.3). Such weighting is better understood in the case of active noninvasive thermal sensing, resulting in the aforementioned feeling that active imaging is more sensitive than radiometry to these extrathermal parameters. But, as a matter of fact, it is possible to show that both approaches suffer from comparable dependencies with respect to these parameters. As a summary, the signal radiated by a body at temperature T measured by a radiometer can be written as:

$$S_{pas} = K_{pas}\{p_i(T),\ p_j\} \cdot T \tag{2.63}$$

while, in the active case, the signal variation resulting from a temperature change ΔT is given, to a first order, by:

$$\Delta S_{act} = K_{act}\{p_i(T),\ p_j\} \cdot \Delta T \tag{2.64}$$

where p_i and p_j denote, respectively, temperature-dependent and -independent parameters.

Similarly, for small temperature changes, the first equation can be written as:

$$\Delta S_{pas} = K_{pas}\{p_i(T):,p_j\} \cdot \Delta T\ . \tag{2.65}$$

From these expressions it clearly appears that active imaging is not very convenient for absolute measurements. But for relative thermometry it can be asked how (1) K (temperature sensitivity) and (2) $\partial K/\partial T$ or $\partial K/\partial p_j$ (temperature accuracy) compare for both modalities. The answers may be given by considering simple configurations for which analytical results can be derived. For plane wave and media configurations, the thermal target can be modeled by means of a thin layer of thickness e, embedded in a slab of total thickness L (Fig. 2.59). The thin layer, located at the depth h, is assumed to differ from the surrounding medium at temperature T only by its temperature $T + \Delta T$. Let $\gamma = \alpha + j\beta$ be the complex propagation constant of the host medium. Its complex index is such that $\gamma = n k_0$, where k_0 is the vacuum propagation constant. Introducing the temperature coefficient c_n, the complex index temperature dependence can be linearized as:

$$n(T + \Delta T) = n(T) + c_n \Delta T\ . \tag{2.66}$$

The complex propagation constant of the thermal target is then given by:

$$\gamma_t = c_n k_0 \Delta T\ . \tag{2.67}$$

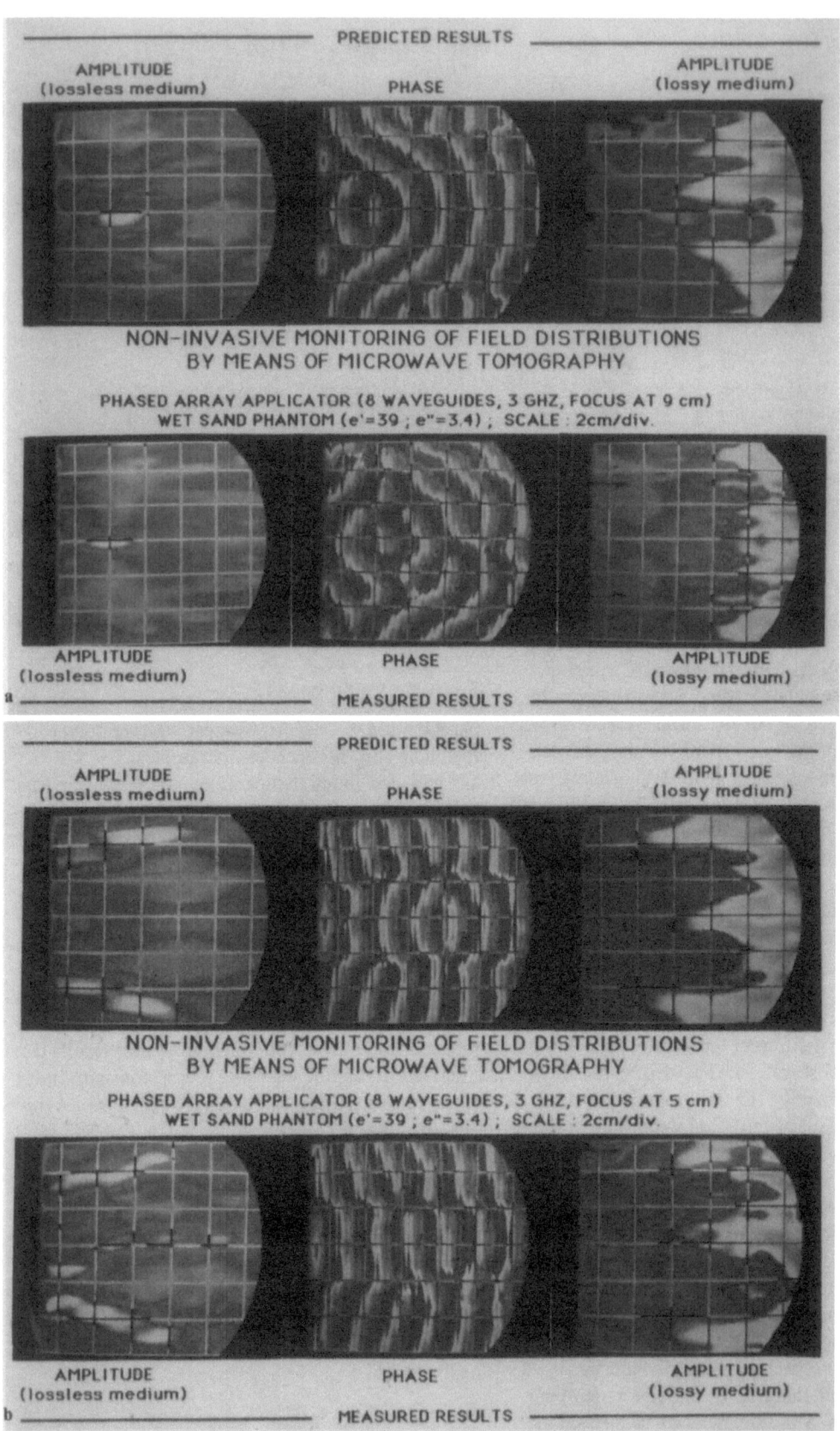

Fig. 2.54a, b. Field distribution radiated in homogeneous phantom simulating lung tissues reconstructed from noninvasive measurements performed by the microwave camera shown in Fig. 2.53. The location of the focusing point according to the amplitude/phase adjustment of the different waveguides is clearly visible on the phase pattern. Comparison with predicted numerical simulation shows relatively good agreement. (Coldefy [29])

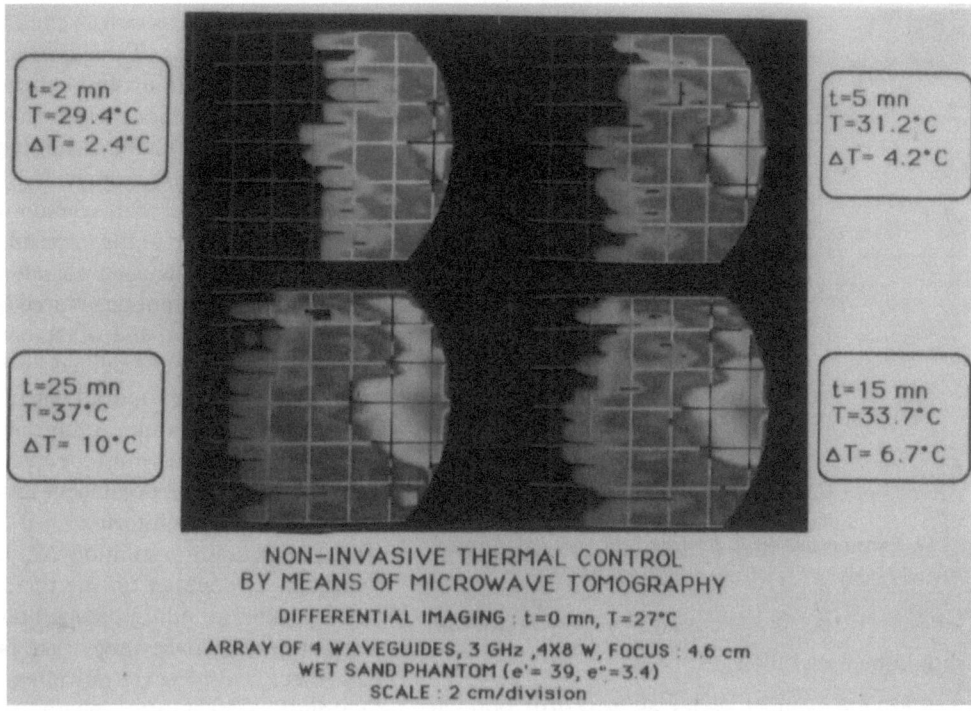

Fig. 2.55. Sequence of differential images when a total power of 8 W is delivered to the phantom. (Coldefy [29])

Concerning the sensitivity aspect, the discussion can be summarized by comparing the $\sqrt{\tau \cdot \Delta T}$ product, where τ is the time constant required to measure the temperature change ΔT. The smaller the value of this product, the better the modality. Simple transmission line calculations give the following results:

$$(\sqrt{\tau \cdot \Delta T})_{pas} = \frac{Q(T+T_R)}{\sqrt{B}} \cdot \frac{\exp(2\,\alpha h)}{\{1-\exp(-2\alpha t)\}} \qquad (2.68)$$

where Q = radiometer constant, T_R = radiometer equivalent noise temperature, B = radiometer bandwidth, and α = attenuation constant, and:

$$(\sqrt{\tau \cdot \Delta T})_{act} = \frac{\{(S/N)\,k_B(T+T_R)\}^{1/2}}{\{\sigma \chi P_{inc}\}^{1/2}} \qquad (2.69)$$

where S/N = signal to noise ratio, k_B = Boltzmann's constant, T_R = receiver equivalent noise temperature, σ_p = absorption cross-section of the probe, and P_{inc} = incident power density, and χ is a constant related to the electrical and geometric parameters of the configuration and modality:

— reflection:
$$\chi_r = |c_n/2n| \cdot |1-\exp(-2j\gamma t)| \cdot \exp(-2\alpha h) \quad (2.70)$$

— transmission:
$$\chi_t = |c_n\gamma_0 t| \cdot \exp(-\alpha L) \ . \qquad (2.71)$$

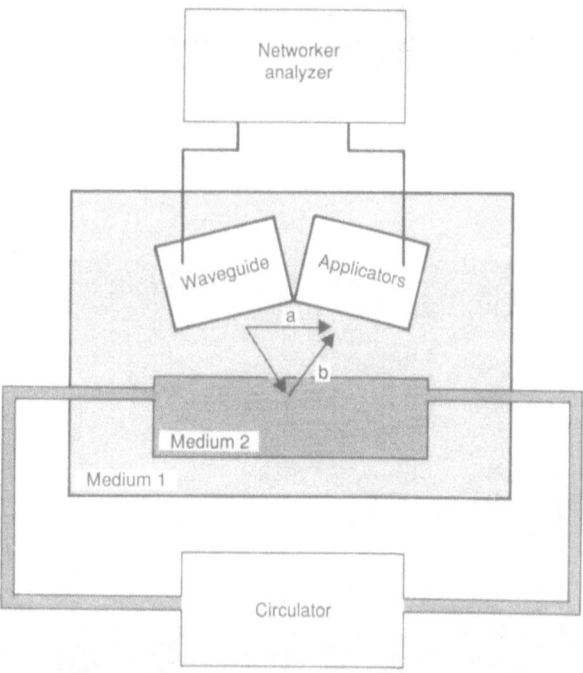

Fig. 2.56. Dual waveguide arrangement combining heating and noninvasive thermal control at 434 MHz. *a* direct path; *b* reflected path. (Hirai et al. [65])

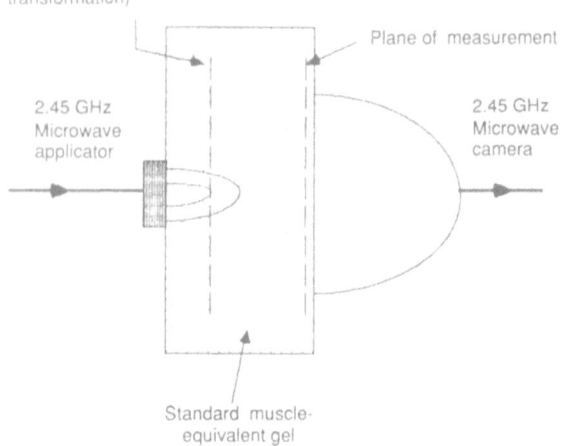

Plane of reconstruction
(near-field to very-near-field
transformation)

Plane of measurement

2.45 GHz
Microwave
applicator

2.45 GHz
Microwave
camera

Standard muscle-
equivalent gel

Fig. 2.57. Experimental setup for near-field probing of hyper-thermia applicator by means of a microwave camera

It should be noted that:

1. The $\sqrt{\tau \cdot \Delta T}$ product varies as $\exp(2\alpha h)$ both in reflection active and passive modalities, and thus is expected to increase rapidly with the depth of the target. Small superficial temperature changes can shadow larger and deeper temperature changes. In contrast, this product is independent of h for active transmission modalities and only depends on the total thickness L of the embedding slab as $\exp(\alpha L)$. L can then be much larger than h for the same value of $\sqrt{\tau \cdot \Delta T}$.

2. It seems that in active imaging the product $\sqrt{\tau \cdot \Delta T}$ can be made as small as desired by means of a suitable increase in the incident power density P_{inc}, but technical as well as clinical considerations limit the usefulness of this possibility. Indeed, the detectability criterion can be dictated by the need to discriminate the measurement signal from parasitic signals more disturbing than thermal noise. Improper probe matching in the reflection case, or imperfect cancellation of the initially transmitted wave in the transmission case, may result in larger parasitic signals than noise and their influence cannot be reduced by decreasing the incident power. On the other hand, the incident power density may be limited by safety considerations, e.g., in order to avoid any thermal effect due to microwave heating. A maximum level of the order of 1 mW/cm^2 must be guaranteed to avoid any damage during permanent irradiation required for continuous monitoring.

3. The temperature resolution ΔT, for a given time constant τ, depends on the thermal target depth and thickness, both in passive and active imaging, demonstrating once more that radiometric measurements must be carefully interpreted in terms of absolute temperature measurement inside the heated tissues (see Sect. 2.3).

For the sake of comparison, the $\sqrt{\tau \cdot \Delta T}$ product has been calculated in a "typical" case at 3 GHz, which corresponds to a frequency already used in existing radiometers. The basic data of this "typical" situation are given in the legend to Fig. 2.60. Inserting these data in previous equations gives:

$$\sqrt{\tau \cdot \Delta T} = 10^{-1} \cdot \exp(0.68 \text{ h}) \quad \text{passive/radiometry}$$

$$\sqrt{\tau \cdot \Delta T} = 8 \cdot 10^{-6} \cdot \exp(0.68 \text{ h}) \quad \text{active/reflection}$$

$$\sqrt{\tau \cdot \Delta T} = 3 \cdot 10^{-6} \cdot \exp(0.34 \text{ L}) \quad \text{active/transmission}$$

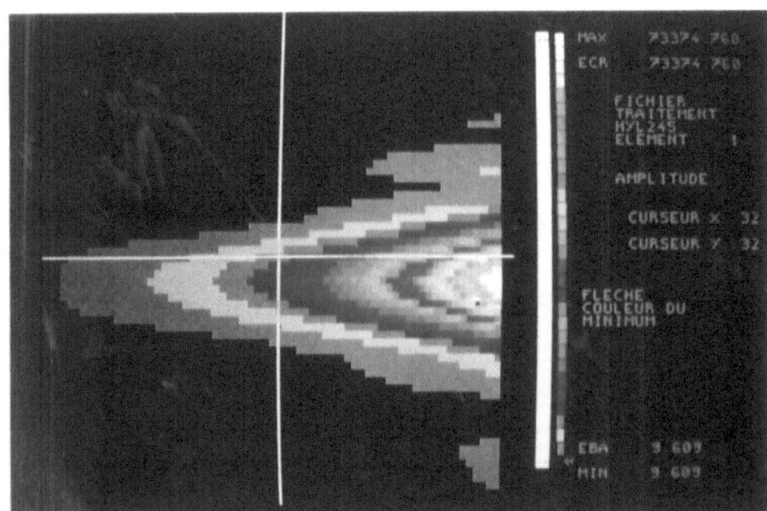

Fig. 2.58. Electric field distribution of a 2.45-GHz applicator reconstructed in a longitudinal plane from noninvasive measurement by means of a microwave camera. (By courtesy of G. Gaboriaud, GBM/TEP procedure)

The main differences between active and passive thermal probing capabilities are clearly illustrated on Fig. 2.60, which displays the variations of the $\sqrt{\tau \cdot \Delta T}$ product with the depth of investigation h. For instance, for a desired temperature resolution $\Delta T = 0.1\,°C$, the measurement rates are, respectively, about 1 s, 10 ms, or 20 ms according to the considered measurement modality.

A complete discussion on the practical usefulness of these results must incorporate dynamic range considerations. Without entering into too much detail, let us only mention that, assuming a -60 dB initial suppression of the reflected signal, the effective investigation depth is limited to approximately 5 cm in the reflection active modality. Due to the plane geometry of the considered model, the investigation depth does not appear to be limited, as long as an initial neutrodyning of -50 dB is realized.

A simple experimental setup has been used in order to confirm the theoretical model. It consists of a 5 cm thick muscle equivalent gel, heated by means of a 434-MHz applicator on its upper face (Fig. 2.61). The thermal sensor − a 3-GHz radiometric probe or a 2.45-GHz microwave camera with 32×32 sensors − is located on the opposite face. The microwave camera is assumed to produce reflection signals on the complex permittivity gradients generated by temperature changes. The temperature can be measured with a thermocouple which is oriented orthogonally to the electric field direction in order to avoid any interfering effect and located at 5 mm depth from the applicator. Figure 2.62 shows a sequence of differential images obtained with the microwave camera operating in the reflection mode. Figure 2.63 allows comparison of the radiometer output as well as the maximum microwave image intensity with the thermocouple reading. It appears that, with the same degree of sensitivity and accuracy, active probing is approximately 1000 times faster than microwave radiometry. During the time in which the radiometer provides one measurement, integrated over a more or less well defined

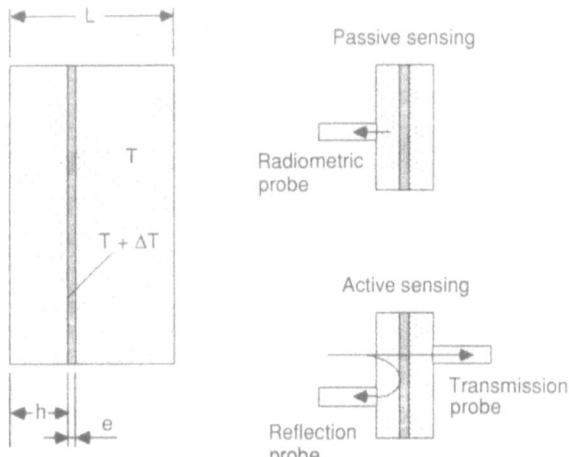

Fig. 2.59. One-dimensional configuration to compare temperature sensitivity of microwave active and passive modalities

volume, the microwave camera gives an image with 5 mm spatial resolution in the transverse plane.

2.6.5.9 Discussion

If all the previous experiments and numerical simulations demonstrate the potential of microwave imaging, it is worth noting that they have been obtained in homogeneous media. The case of inhomogeneous media appears more problematic.

Experiments have been conducted on isolated horse kidneys. Using multiview procedures, differential tomographic reconstructions have been achieved when the temperature of the perfusing liquid was changed [50]. Although image changes have not been correlated to temperature variations, the main result is that microwave images are sensitive to temperature changes of only a few degrees.

Concerning numerical modeling, some good results have been obtained in the case of a relatively simple structure like a human arm [87]. Multiview imaging

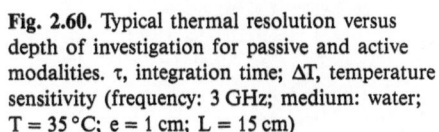

Fig. 2.60. Typical thermal resolution versus depth of investigation for passive and active modalities. τ, integration time; ΔT, temperature sensitivity (frequency: 3 GHz; medium: water; $T = 35\,°C$; $e = 1$ cm; $L = 15$ cm)

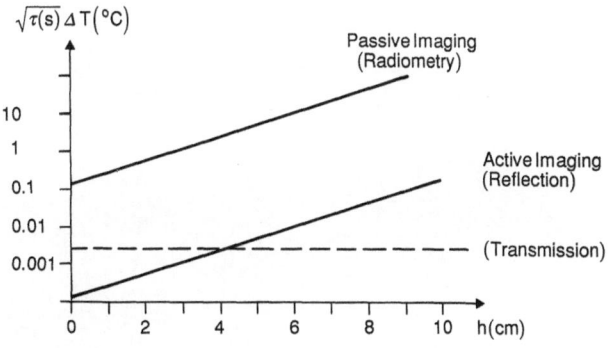

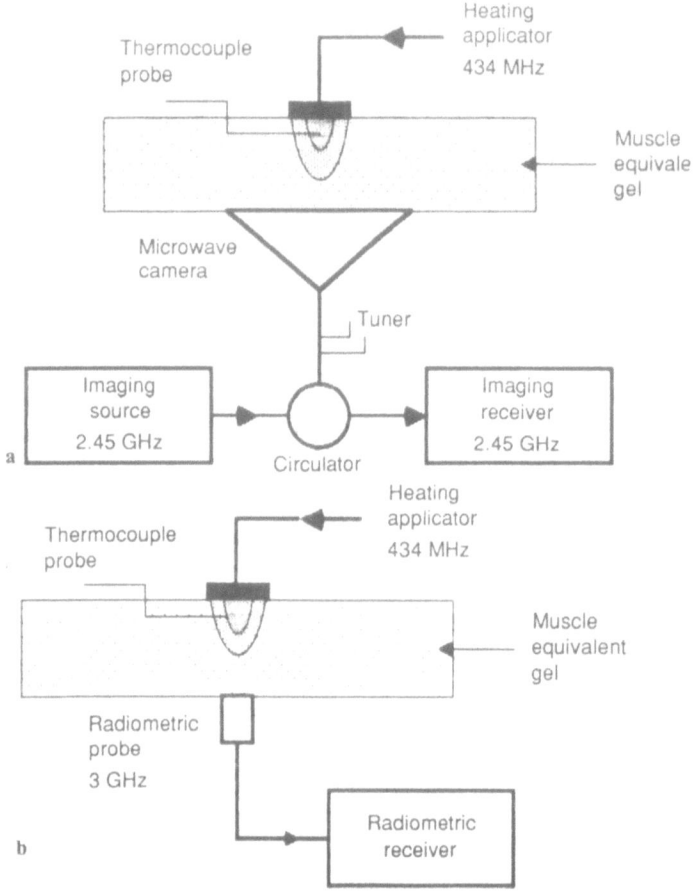

Fig. 2.61. Experimental setups to compare **a** active and **b** passive microwave thermal sensing capabilities

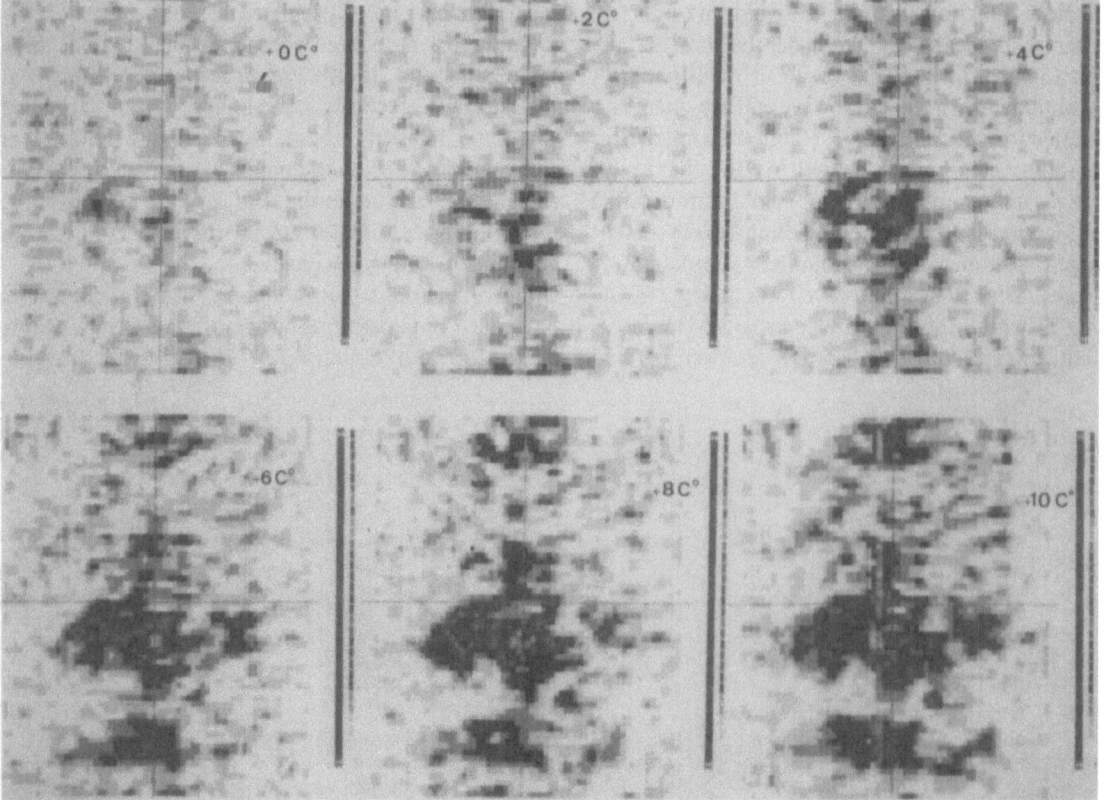

Fig. 2.62. Active microwave images in reflection mode for microwave radiative heating of a gel block

allows a satisfactory reconstruction of the different media: fat layer, muscle, and bone. Furthermore, a small temperature increase inside muscle is conveniently detected and localized through differential imaging. In the case of a more complex structure like a neck (see Fig. 2.47), numerical simulations provide bad results concerning the reconstruction of the different anatomical parts [34]. Although better results are obtained with differential imaging to visualize local temperature changes, some artifacts can appear in some configurations, for instance when the temperature change is localized near strongly diffracting parts like bones. Such results demonstrate the inadequacy of presently available algorithms and the need for further research in diffraction tomography algorithms.

2.6.6 Radiofrequency Inverse Scattering Techniques

2.6.6.1 Extension of the Imagery Concept

Previous results have been obtained by means of a specific inversion schema for a given geometry. That is to say that in the geometries under consideration, the general equations relating the scattered field to the equivalent current can be more or less accurately managed by introducing well known mathematical transformations – Fourier, Hankel, etc., – which can be easily handled from a numerical point of view.

But, these basic relationships could also be considered from a more general point of view (Fig. 2.64). Instead of the points of measurement of the scattered field having to be located on a plane or a cylinder – as in optics or in X-ray technology –, these sampling points could be, at least in principle, located "anywhere" in the space surrounding the body under con-

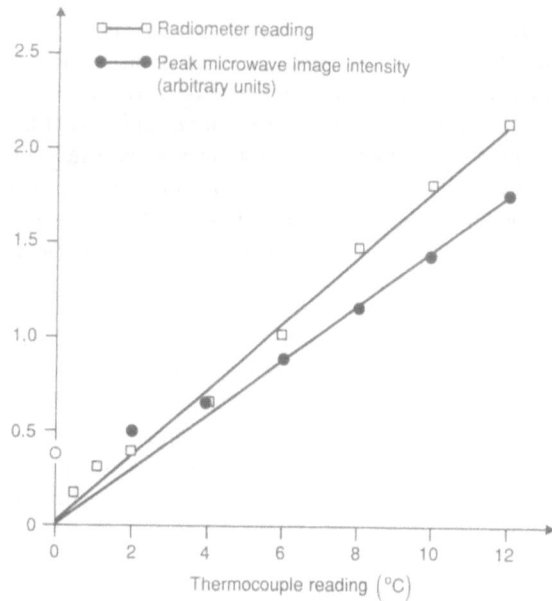

Fig. 2.63. Correlation of active and passive measurements with temperature

sideration. Such an arbitrary location offers more flexibility for possible arrangements, but, conversely leads to loss of the simplicity of calculation associated with more conventional geometries. The aforementioned flexibility involves a better arrangement of the probes around the target according to the available space and the surface to be covered, but it also applies to the source location and the selection of the operating frequency in more favorable domains with respect to attenuation in living tissues. With regard to this aspect, radiofrequencies would probably provide a suitable frequency range.

The most intriguing feature of such imaging techniques is that the spatial resolution does not seem to be limited by wavelength considerations as in diffracted limited imaging systems. Starting from an initial meshing of the target, it only appears necessary to

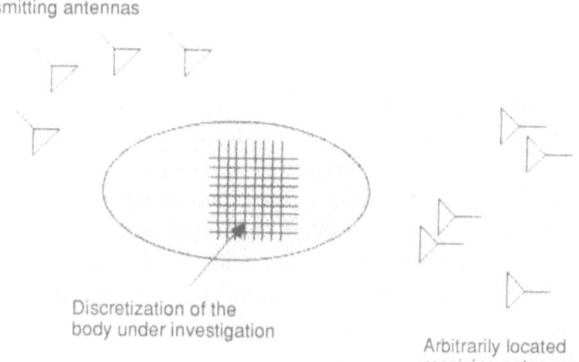

Fig. 2.64. Schematic arrangement for generalized imagery

make the sufficient number of measurements to determine the desired quantity in each, two- or three-dimensional, cell. Elementary considerations of information theory indicate that the number of items of independent information cannot be greater than the number of independent measurements. If N a priori independent cells are considered, at least N measurements have to be achieved. This means that, in practice, the independence of real, noise-corrupted, measurements is difficult to ascertain.

As a matter of fact, the numerical task consists in inverting one or more linear systems of equations which are more or less stable with respect to noise. Such a generalized approach to the imaging and tomographic problems is now being investigated from a theoretical point of view, and preliminary numerical results illustrate its potentials and limitations.

2.6.6.2 Algorithms for Generalized Imaging

As previously explained, diffraction tomography provides, in the best cases, the induced equivalent current distribution $J(x)$, mixing local dielectric properties $\varepsilon(x)$ and local total field dependence $E_t(x)$. Several algorithms have been proposed to remove the field dependence in order to isolate dielectric property distributions. All of them are based on the electric field integral equation. Two typical examples will be briefly described.

The first consists of three steps. The linear relationship existing between the scattered field $E_s(x)$ and $J(x)$ allows the determination of $J(x)$, inside the object, from the values of E_s measured at p distinct locations x. Knowledge of $J(x)$ allows, in its turn, the calculation of the total field $E_t(x)$ everywhere and, especially, in the object. Finally, the simultaneous knowledge of the internal total field $E_t(x)$ and $J(x)$ leads to the determination of $\varepsilon(x)$. Simple in its principle, this first algorithm was found to be quite unstable and very sensitive to the measurement errors on the scattered field.

A second algorithm is based on a prior arrangement of the electric field integral equation before handling numerical inversions [46, 55]. This arrangement consists of a convenient a priori separation between the intrinsic dielectric properties of the object and the total field. For this purpose, the object is assumed to be divided into a large number of elementary cells, while transmitting and receiving antennas are disposed around it. For the sake of simplicity, the scattered field will be formally written as a function of the total internal field $E_{t,1}(x)$ as follows:

$$[E_{s,2}] = [B] \cdot [E_{t,1}] \ . \tag{2.72}$$

In the above equation, the indexes 1 and 2 relate, respectively, to the internal and the external parts of the object. The first matrix could result from the measurement of the three components of the scattered field at N points, in such a way that $[B]$ represents a $3N \times 3N$ matrix. The internal total field $E_{t,1}$ can be formally expressed versus the internal incident field $E_{i,1}$, via a classical integral equation (moment method), as:

$$[A] \cdot [E_{t,1}] = -[E_{i,1}] \ . \tag{2.73}$$

The matrixes $[A]$ and $[B]$ can be rearranged in this way:

$$[A] = [A_1] \cdot [R] - [I] \ , \tag{2.74}$$

$$[B] = [B_1] \cdot [R] \tag{2.75}$$

where $[I]$ is the identity matrix, $[A_1]$ and $[B_1]$ only depend on the experimental conditions (frequency, discretization cell of the object, measurement locations), and $[R]$ is simply a diagonal matrix whose elements are equal to the permittivity contrast $\varepsilon(x) - \varepsilon_a$ at the considered cell location. Such a factorization to separate experimental parameters from the dielectric properties of the object is the key to this second algorithm. Some trivial, but lengthy, matrix calculations allow the determination of the following quantities:

1. Complex permittivity
2. Total electric field (amplitude and phase)
3. Equivalent current
4. Rate of energy deposition

Put briefly, this algorithm should provide all quantities of practical interest for non-invasive control of hyperthermia. In practice, however, its efficacy is limited by stability problems, as is usual in solving inverse diffraction problems which are known to be ill-conditioned. Nevertheless, the generality of the experimental arrangement, the available flexibility in choosing the frequency, the measurement points, the transmitting antenna locations, the embedding medium, etc., and the possible use of a priori information on the object offer better conditions for attacking the conditioning difficulties.

Other approaches, more or less related to the previous one, are also being investigated [51].

2.6.6.3 Preliminary Numerical Simulations

Two- and three-dimensional calculations have been performed to test the feasibility of generalized imag-

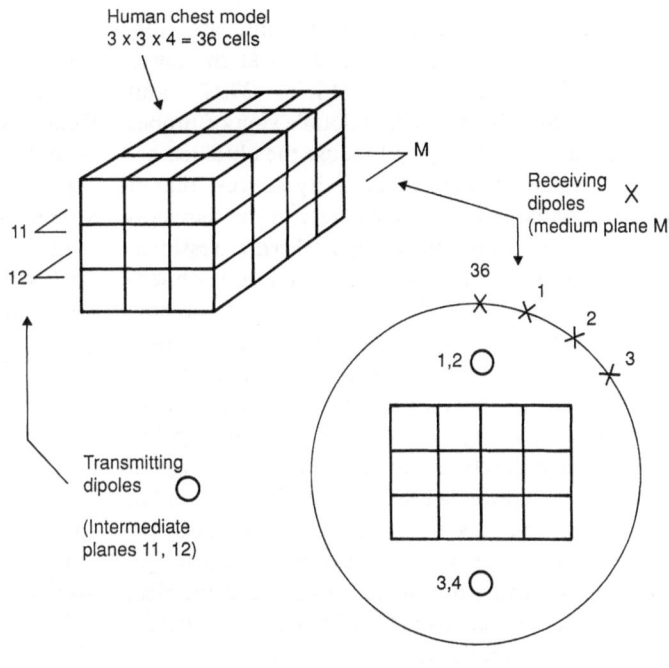

Fig. 2.65. Cubic human chest model and antenna arrangement for generalized imagery. (Ghoadonkar et al. [46])

ing algorithms. The measured data were simulated by numerical means, using classical determination of the induced equivalent currents by means of the electric field integral equation (moment method) and further calculation of the scattered field at the probe locations. Measurement errors were introduced by adding random numbers to the previous exact results. It was thus possible to estimate the stability of the algorithms with respect to data noise corruption. Typically, 0.4% relative amplitude and 0.05° phase errors were introduced.

Among the considered three-dimensional geometries, a cubic, three-layer, 36-cell body has been used to simulate the human chest cavity [46]. The cell size was 7.5 cm, or, equivalently $\lambda_0/25$ at the operating frequency of 150 MHz, providing reasonable resolution and significant depth of penetration. The complex permittivity of each of the cells was adjusted by taking into account the mean value of all its constituents according to their volume ratio. The model dimensions were $30 \times 22.5 \times 22.5$ cm, while probes were located on a circle of radius equal to 17.5 cm and centered on the model. Four transmitter locations were considered (Fig. 2.65). The distance between the 2 cm high transmitting dipoles and the model was greater than twice their height, which proved to be sufficient to neglect any interaction between dipoles and model.

As a significant result, it appeared that the error on the complex permittivity retrieval was 100 – 300 times

the input error, illustrating the extreme instability of the inversion scheme. Acceptable relative errors of the order of 2% were, however, obtained by assuming the model to be immersed in a 0.58% saline solution. For the sake of comparison, the relative errors were approximately 70% of those for the same model in air. Indeed, a suitable immersing medium reduces reflection losses and improves wave penetration in the model, resulting in a better sensitivity in the perception of its interaction with the model. Additional tests were conducted in order to evaluate the minimum height of the model for performing satisfactory permittivity reconstructions in a given plane. Only a few differences appeared when the number of layers was increased from 4 to 6 and to 24, in the case of 8 cells per layer. However, the accuracy of reconstruction rapidly decreases when the number of fully buried cells – i.e., cells with no face at the periphery of the model – is increased.

Although very preliminary, these typical results clearly illustrate the limitations of this approach in terms of model complexity. Until more powerful computers are developed, the number of cells will remain severely limited. This is the reason why two-dimensional calculations have been conducted in order to study more systematically the effects of different experimental and numerical factors in less memory-consuming situations. Muscle homogeneous or bone/ muscle inhomogeneous models, a single or double ring of measurement probes, various immersing media, multi-illumination configurations, and different algorithms have been considered at the lower hyperthermia frequencies of 13.56, 27.12, and 40.68 MHz. Simulations with a double ring of probes are particularly encouraging because the obtained accuracies in the complex permittivity suggest that it may be possible to calculate changes in the temperature and/or perfusion during hyperthermia sessions. More recent theoretical developments on three-dimensional problems show that, by means of suitable algorithms, the use of multiview procedures permits matrix inversion to be limited to a single matrix of rank $\sqrt{N}$, N being the total number of cells. As a consequence, a probing array of 200 fixed dipoles, operating successively as emitters and receivers, would allow consideration of a total number of cells equal to 360000, instead of 200 with previously described procedures. The geometry of the experimental arrangement appears to be not so far from the cylindrical tomographic configuration considered at the beginning of this section. But the algorithms consist of very different numerical processing whose efficiency still has to be demonstrated from numerical and practical points of view.

2.6.6.4 Optimization of Hyperthermia Treatments

The optimization of the hyperthermia treatments is the next step after achieving their noninvasive (or, evidently, invasive) control. Using a set of applicators and sensors, the problem consists in employing the data provided by the sensors to adjust the amplitude and the phase (at least in the coherent case) of the waves radiated by the applicators. A number of papers have been devoted to the possibilities offered by arrays of electrodes or applicators to focus or concentrate the power deposition at the desired location.

The linearity of the relationship between the internal local field and the applied external sources is at the foundation of such optimization approaches. Here again, the numerical techniques which are effectively used differ from one case to another. In so-called direct problems, the unknown is the internal field or current distribution in a well-defined structure resulting from know applied sources − or, equivalently, applied fields. The same linear equation can be used to solve the inverse problem, consisting of finding what external sources would be able to induce a given internal field or current distributions in the same structure.

For instance, RF heating at 20 MHz has been investigated by means of two arrays of electrodes located on both faces of a three-dimensional, 9 cm × 6 cm inhomogeneous model, including fat layer and muscle core [64]. Due to symmetry, only one-eighth of the model is discretized into 144 1.5 cm volume cells. Each array consists of 432 subelectrodes, almost covering the model surface. The prescribed internal field distribution was assumed to be constrained as follows:

1. Polarized perpendicularly to electrodes
2. Maximum in the muscle region
3. Minimum in the fat layer
4. Zero outside the volume internal to the electrodes

Such optimal behavior was found to be created by potentials on the subelectrode exhibiting very high oscillating values, difficult to apply effectively in realistic situations due to coupling between adjacent subelectrodes. Still more unrealistic and unstable are the potentials desirable for a given SAR distribution.

Another illustrative example is provided by microwave heating of chest regions by means of an array of 50 small sources at 915 MHz [4]. In this case, instead of directly attacking the inverse problem by a single matrix inversion, an optimization technique has been used to obtain the feeding conditions of the antennas in order to minimize power deposition outside a given area. The model is two-dimensional but includes seven different media identified from CT scans. The antennas are consequently line sources and are located around the body cross-section. They are assumed to be uncoupled and the reaction of the body is neglected. Both phase optimization (equal amplitude for all sources) and amplitude/phase optimizations have been considered. Such an approach seems preferable with respect to stability of results, especially in the presence of small errors of the assumed complex permittivities of the different tissues. Amplitude/phase optimization provides better concentration while reducing the total power delivered to the body. Indeed, eight sources out of the 50 have practically negligible contributions.

These two selected examples, even if they foster two different feelings as to the possible success of multiparameter heating equipment, pose the crucial problem of the adjustment of these parameters in clinical situations. These adjustments must be achieved before and during heating and need adequate criteria. The necessity of continuous monitoring results from the fact that, if power deposition capabilities at a given point are a necessary condition to achieve hyperthermia, complex and time-varying thermoregulatory mechanisms may cancel the advantages of the initial adjustment during the treatment. Generalized imaging techniques seem to offer the most convenient flexibility to achieve noninvasive measurements from which adjustment criteria will be deduced. Unfortunately, the corresponding numerical tools are not yet available.

2.6.7 Conclusion

The salient features of dielectric imaging can be summarized as follows:

1. The complex permittivity is a sensitive indicator for direct and indirect effects of temperature changes, even in depth applications.
2. Convenient equipment exists for rapid and accurate collection of a lot of data.
3. The general relationship connecting the required unknown quantities (complex permittivity, internal field or SAR distributions, etc.) to the known data (scattered fields or perturbed voltages/currents) is given by closed form expressions, even if, theoretically, existence and uniqueness aspects are not as clearly felt as for direct problems.
4. Simple approaches extended from quasi-optical or quasi-static approximations do not perform very conveniently, i.e., quantitatively, the inversion of the considered non-linear problem, unless under restrictive conditions. Nevertheless, in clinical situations they are already able to provide useful qualitative information that is otherwise inaccessible.
5. Existing numerical processing for generalized imaging is not yet really convenient for extracting the desired quantities from the measured data (insufficient stability, long computation time, large memory requirements, etc.).
6. There is an evident need for basic living tissue dielectric characterization and development of stable algorithms.

2.7 Ultrasonic Techniques

2.7.1 Presentation

Like X-rays, ultrasonic waves have been used in medical imaging for a long time. For this reason, many forms of ultrasonic equipment are in use for various diagnostic purposes. Echotomography is probably one of the most popular ultrasonic modalities to be used in order to noninvasively investigate soft, muscle-like tissues. Among its recognized advantages, ultrasonic imaging provides innocuity, satisfactory spatial resolution and collimating capabilities resulting from the small wavelengths at the operating frequencies. For frequencies of the order of a few megahertz, the wavelength is smaller than 1 mm. Furthermore, the relatively low propagation velocity, of the order of 1500 m/s in soft tissues, allows easy time discrimination of the echoes coming from discontinuities crossed by the incident beam. Radar technique may be used for short-range target situations, in contrast to the microwave case, which suffers from much higher propagation velocity. For all these reasons, ultrasonic techniques rapidly appeared as good candidates for noninvasive control of hyperthermia. But the success of these techniques for this purpose depends on the sensitivity of the propagation conditions of sound in living tissues with respect to temperature. This aspect, which constitutes one of the main difficulties in developing ultrasonic noninvasive means of control, is examined in the next section. Subsequent sections are devoted to the possible approaches, including both active and passive modalities, as in the case of microwaves. In addition, so-called thermo-induced acoustic imaging is also considered, consisting of exciting acoustic noise by proper stimulation sources such as RF or microwave pulses and ionizing beams. Finally, the present state of the art is discussed and some extrapolations drawn from the few available results.

2.7.2 Sensitivity of Living Tissue Characteristics to Temperature

As for microwaves, the propagation properties of tissues can be described by their plane wave propagation constant. As a matter of fact, both sound velocity v and attenuation α are temperature dependent, opening the way to the possible use of amplitude and phase processing of the signal transmitted by heated tissues. For example, in the case of soft tissues, which constitute the most propitious propagating medium for ultrasonic waves, the velocity dependence is very close to that in water, which is approximately given by:

$$v = 1403 + 5\,T + \text{higher order terms} \qquad (2.76)$$

where v is in meters per second, and T in degrees Celsius. In vitro measurements on soft tissues around

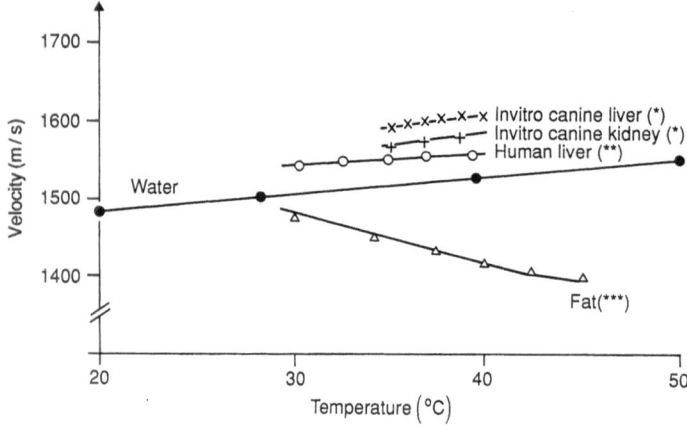

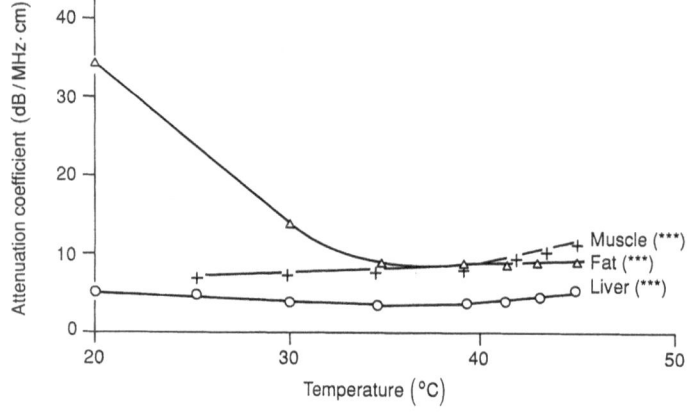

Fig. 2.66. Typical variations of sound velocity and attenuation constant in different tissues: (*) Nasoni et al. [80]; (**) Rajagopalan et al. [86]; (***) Aida et al. [2]

40 °C have shown temperature sensitivities of the order of 0.5 – 4 m/s/°C and have confirmed that velocity is an increasing function of temperature, as illustrated in Fig. 2.66, which summarizes some typical results obtained by several groups [2, 10, 66, 83, 89]. Although the absolute value of the velocity is not constant, the slope of the velocity versus temperature appears to be reasonably uniform.

The case of fatty tissues is less clear. For these tissues, the velocity can decrease with frequency. Similarly, depending on the tissue nature, the temperature, and the frequency ranges, the attenuation constant change may be positive or negative. Attenuation constant measurements are less accurate than velocity measurements. However, signal amplitude and spectral difference methods have provided comparable results, as shown by Benjapolakul et al. [10]. In vitro measurement reveals temperature sensitivities of the attenuation constant of the order of ±1.5%/°C.

Such variation in the behavior of the living tissues with respect to temperature points out a real need for in vivo quantitative characterization.

2.7.3 Active Modalities

The first attempts to use the velocity dependence with respect to temperature for noninvasive thermal sensing have been devoted to single path measurements. Single path procedure means that the variation of the sound velocity is determined along the expected beam trajectory. Both reflection and transmission modes have been investigated. The localization of the temperature elevation in the tissues on the beam trajectory can be achieved by a convenient analysis of the incident pulses reflected from the tissue discontinuities resulting from their temperature elevation. Transmission techniques do not provide such localization capabilities, as far as single path measurements are involved.

For instance, early results have been reported by Robert et al. [91], using an Aloka SSD-30 echograph at the operating frequency of 2.25 MHz. The experimental schema is illustrated in Figure 2.67. A 1- to 3-MHz power generator is used to heat a phantom consisting of a small glycerol target immersed in a low

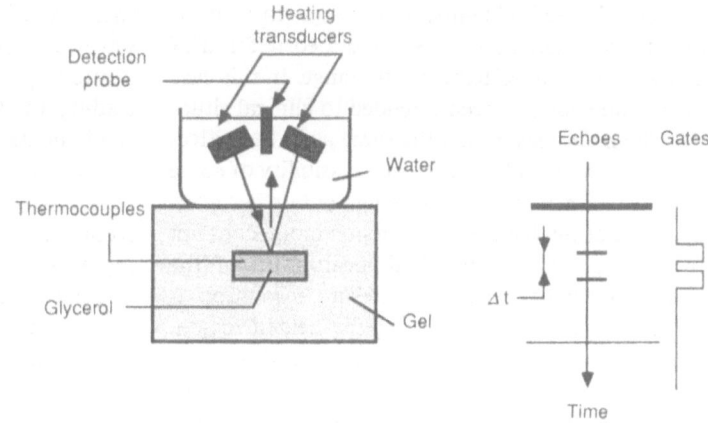

Fig. 2.67. Noninvasive temperature measurement by means of ultrasound echography (Robert et al. [91])

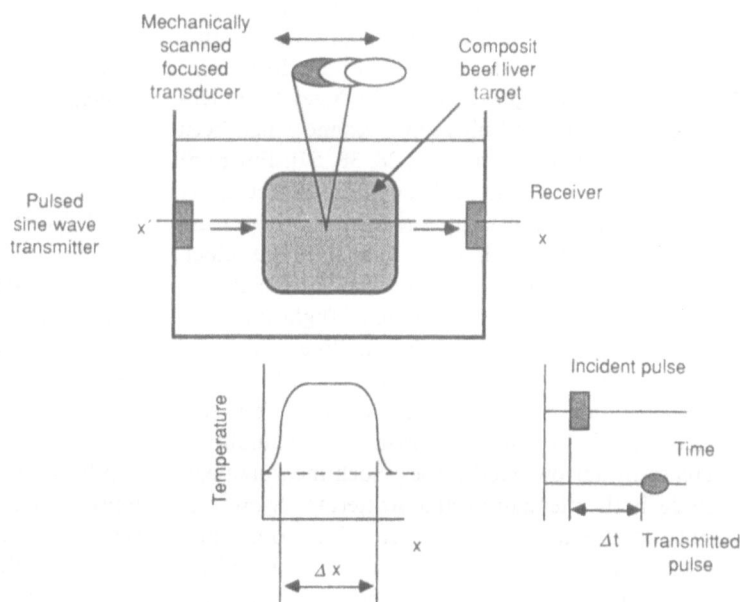

Fig. 2.68. Noninvasive temperature measurement through transit time of transmitted ultrasound pulses (Davis and Lele [32])

absorbing gel. An electronic gating system allows the selection of the echoes coming from the identified glygerol/gel discontinuities. Knowledge of the temperature coefficient of the sound velocity in glycerol (approximately 1.15% per 10 °C) allows the translation of the time delay between the two selected pulses to the temperature increase in the glycerol sample. The accuracy and the stability of the time delay measurements should provide relative temperature precision of the order of 0.5°–1.5 °C at 5–6 cm in depth and within 10 s. Comparative temperature measurements with thermocouples show relatively good agreement. It is, however, clear that velocimetry does not measure the temperature at a given point, but an averaged temperature between the two selected interfaces. Promising in vitro results have not been obtained in vivo: the absence of well-defined boundaries, patient move-

ments, and the great variability of the temperature sensitivity with temperature explain the difficulties encountered.

Transmission mode has been considered by Davis and Lele [32]. The phase shift of an interrogating pulsed beam is used to measure changes in its propagation velocity through the heated region. The experimental setup is shown in Fig. 2.68. Experiments were conducted by Davis and Lele in vitro and in vivo using a diagnostic frequency of 2 MHz and a heating frequency of 2.7 MHz. The heat deposition spot was achieved by means of a mechanical scanning of a focused beam. The length of the heated region was approximately 3 cm. The phantom consisted of a layered composite of beef and liver tissue. By using adequate averaging processes, the correlation between change in propagation delay and temperature was found to be

0.99 over a 21°–30 °C temperature range. In vivo experiments conducted on rats showed a correlation of 0.92, over a 32°–40 °C temperature range. In this case the technique has not been extended to clinical situations. The main reasons are the same as in the reflection case, with, in addition, the inaccessibility of some tumors for transit time measurements. Furthermore, it is worth noting that the transmission mode does not allow for accurate longitudinal localization of the temperature change. The temperature resolution is then dependent on the length of the heated region. For example, with the described experimental setup, the resolution is approximately 2 °C and 0.2 °C for lengths equal to 1 cm and 10 cm, respectively. The localization of the temperature elevation is expected to occur in the focusing area of the heating transducer, which needs to be known a priori.

Following the same guidelines as for X-ray tomodensitometry, the idea of ultrasonic scanners has been considered for a long time [e.g., 28, 38, 57]. Pulses of ultrasound are transmitted through the object under investigation. Time of flight and frequency spectrum measurements are used in order to analyze the velocity and the attenuation constant. Several techniques have been tested. Concerning time of flight measurements, the time delay represents the line integral of the transit time of the pulse through its path. For the attenuation constant, three methods have been suggested. The first consists in measuring the amplitude decrease of narrow band pulses. Such measurements include both attenuation and scattering. More sensitivity to attenuation was expected from other procedures which consist in forming the ratio of the received to the transmitted energies at the mean frequency, or observing the shift of the median frequency of the received pulse compared to a reference pulse transmitted in a reference medium such as water (time domain spectroscopy). The main difficulties with

such procedures result from the fact that ultrasonic waves do not propagate as linearly as X-rays.

Scattering and diffraction effects predominate, leading to diffraction tomography algorithms. The fundamental difficulties are the same as those encountered with microwaves. In addition, technological problems are more critical, resulting from the smaller wavelength. The problem of realizing arrays of sensors with a spacing of the order of one-half of a wavelength does not seem to have been conveniently solved. Furthermore, considerations concerning accessibility for shadowed regions limit the range of applicability of tomographic processes. Finally, although many efforts seem to have been devoted to tomographic approaches, sometimes with good results at the numerical modeling level, no significant progress has emerged at the practical or clinical level.

2.7.4 Ultrasound Radiometry

Considerations applying to microwave radiometry (see Sect. 2.3) can be used in ultrasonic radiometry [28]. Two main differences concern the wavelength and the attenuation constant. From both points of view, ultrasonic radiometry could be expected to provide better spatial resolution and greater penetration depth. Furthermore, due to Planck's formula, the noise power density, which varies approximately as the inverse of the square of the wavelength, should be larger and the measurement more sensitive.

While lateral spatial resolution can be obtained by means of a transducer with moderate size, the axial resolution could be attained with cross-correlation radiometers as shown on Fig. 2.69. For a given time delay between two transducers located at opposite positions, it is possible to isolate the contribution from that volume which is equidistant in time from both receivers. By varying the time delay, the position of the volume under investigation can be moved along the beam axis between both transducers. This cross-correlation technique is claimed to be insensitive to tissue absorption coefficients located between the investigated volume and the transducers. Despite attractive theoretical limits on performance, ultrasonic radiometers have not yet been used clinically.

2.7.5 Thermo-induced Acoustic Imaging

When a medium is illuminated by an incident radiation pulse, a part of the incident power is absorbed.

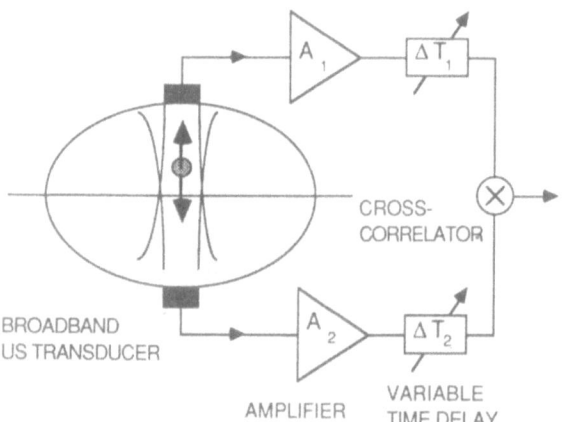

Fig. 2.69. Ultrasound correlation radiometer. (Christensen [28])

The resulting sudden thermal stress causes a rapid and very small rise in temperature, of the order of $10^{-5\circ}$ or $10^{-6\circ}$C. The corresponding thermal expansion produces a strain in the medium and generates an ultrasonic wave that propagates away from the illuminated region. This wave can be detected by usual means and, for example, the contributions of the different parts of the medium can be separated by usual time of flight techniques.

The incident radiation can be ionizing (X-rays, high-energy electrons) or not (radiofrequencies, microwaves, ultrasound) [16]. The thermoacoustic wave generation depends both on the local characteristics of the tissues and on the nature of the incident pulse. It is worth noting that the expected dynamic range, in otherwise comparable situations, is five times larger in the case of radiofrequencies or microwave pulses because of the wider range of variations of the conductivity of biological tissues.

According to the incident pulse duration, the spectrum of the acoustic signal extends from a few KHz to a few MHz, with a maximum close to 1 MHz. For microwave illumination, the spectrum of the ultrasonic pulse can be written as follows [72]:

$$|F(j\omega)| = \frac{\sqrt{2}}{\omega}\{1 - \cos(\omega\tau)\}^{1/2} I \cdot \frac{3\beta v}{c} \cdot \frac{2\alpha v\omega}{\omega^2 + 4\alpha^2 v^2}$$

(2.77)

where the different symbols and related typical values are indicated in Table 2.5. It can be shown that the first term corresponds to the Fourier transform of the illuminating microwave pulse with energy $I\tau$; the second term holds for the thermoelastic conversion and the third one describes the frequency dependence involving material properties such as sound velocity and microwave attenuation constant. The proportionality between the intensity of the acoustic signal and the illumination power density can be used to display the spatial power density profile and hence the spatial distribution of the rate of temperature rise.

While the signal strength has been shown to be proportional to $I\tau$, a more complete calculation would exhibit a τ^2 dependence of the signal to noise ratio [16]. In order to obtain convenient spatial resolution, the pulse duration is limited to approximately 1 μs. The spatial resolution could be increased while maintaining τ constant by increasing the pulse peak power level I, at the possible risk of cell damage of membrane breakdown.

Several experiments have been conducted, using different kinds of illuminations (capacitive RF electrodes, microwave horn antennas or open cables, etc.) on various configurations (small spheres or cylinders, simulated human hand or head, etc.) [23, 72, 72, 85,

Table 2.5. Physical quantities and related symbols relevant to thermo-induced imaging

Symbol	Quantity	Typical value	Unit
τ	Microwave pulse width	0.5	μs
I	Microwave peak power density	1000	W/cm^2
ω	Ultrasonic frequency	1	MHz
β	Thermal expansion coefficient (water)	0.0001	K^{-1}
v	Sound velocity (water)	1500	m/s
c	Specific heat (water)	4.2	J kg^{-1}K^{-1}
κ	Thermal conductivity coefficient (water)	0.58	W m^{-1}K^{-1}
α	Microwave attenuation constant (water, 3 GHz)	62	m^{-1}

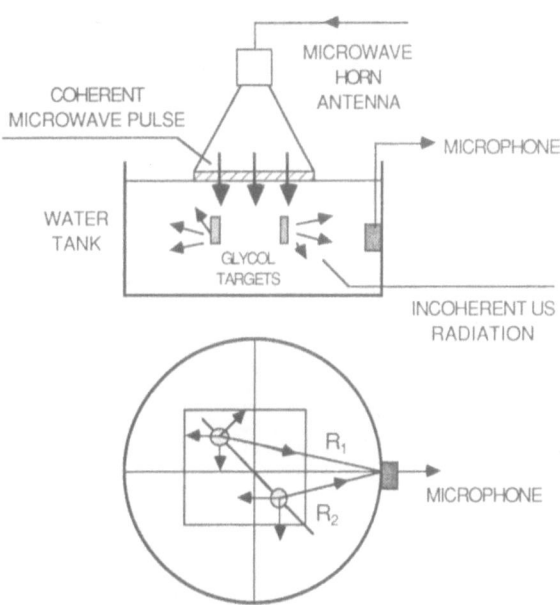

Fig. 2.70. Experimental arrangement for thermo-induced acoustic imaging. (Krug et al. [72])

86]. These experiments have confirmed the feasibility of thermo-induced acoustic imaging, but the question as to whether the sensitivity is sufficient for noninvasive thermometry purposes is still open.

A recent example is illustrated by Fig. 2.70 [72]. The incident radiation consists of a 3-GHz microwave pulse illuminating a water tank with small cylindrical or spherical targets (20×25 mm). Biological tissues of high and low water content have been simulated by water and glycol, whose thermoelastic conversion factors are 0.11 mN/W and 1.28 mN/W, respectively. The peak microwave power density is 1000 W/cm^2 during the pulse duration $\tau = 0.5$ μs. The power is

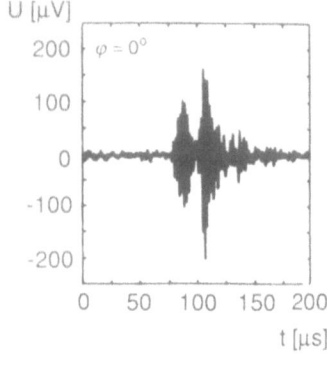

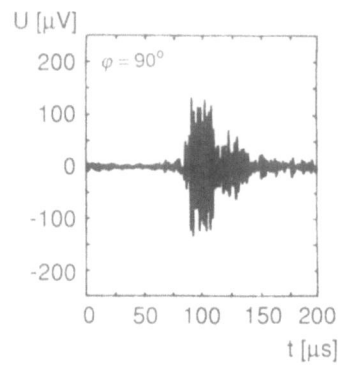

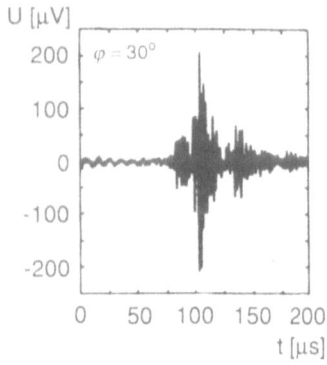

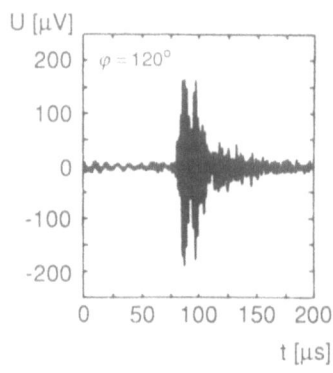

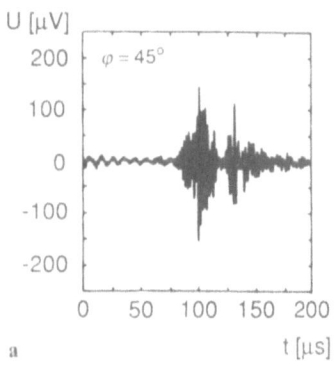

2-Target
responses

a

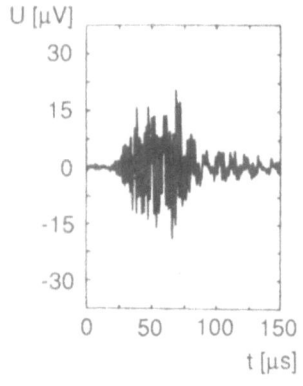

Without target

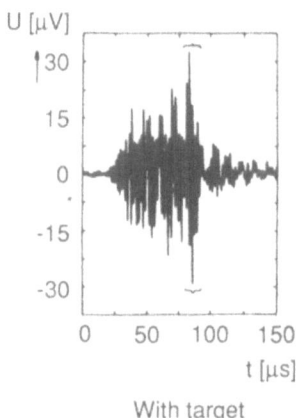

With target

b

Fig. 2.71. Ultrasound signal response at 0.5 MHz related to the arrangement shown in Fig. 2.70 (by courtesy of J. Krug and P. Edenhofer): **a** Echoes from small targets (cylinder 25 mm ∅; sphere 35 mm ∅), according to their mutual orientation with respect to the microphone. **b** Sensitivity of the measurement estimated by the smallest object detectable (cylinder 3 mm ∅) from noise and parasitic signals in the water tank without in-homogeneities

radiated by means of a standard S-Band horn which is matched to water with a quarter wavelength slab. The targets are located at 10 mm from the matching slab. The ultrasonic signal is received by means of a microphone.

According to the position of the targets with respect to the microphone, the echoes coming from the targets can be separated in time with an apparent spatial resolution of approximately 10 mm, and probably with a better one by using more refined data processing. By comparison to residual signals resulting from noise or parasitic reflections from the water tank, the sensitivity is estimated by the smallest detectable diameter of target, in this case approximately 3 mm (Fig. 2.71). It is not so easy to translate such a sensitivity into a temperature change during the hyperthermia treatment.

As far as the rise and fall time of the microwave pulse are negligible before its width τ, the intensity of the ultrasonic wave varies linearly with the microwave peak power over more than one decade, between 50 and 1400 W/cm^2. Some authors indicate that, with regard to thermal dose, only the mean power has to be considered. Such peak powers would seem to be compatible with usual standards. However, it is not really clear whether incident power densities present no risk of cell damage by electric breakdown of membranes. Probably higher power could be tolerated with ultrasonic pulses before the advent of lesions due to cavitation. But ultrasonic illumination is not very convenient for producing uniform heating owing to low sound velocity, and, furthermore, the incident pulse would provide direct scattering from which the thermo-induced signal would have to be extracted.

As a matter of fact, such an approach seems very attractive in the sense that it presents the unique feature of combining a satisfactory uniformity of the illumination at large depths (for instance, by using X-rays or electromagnetic waves at sufficiently low frequencies) and, independently, a good spatial resolution provided by the thermo-induced acoustic signal. However, the limitations already mentioned for classical ultrasonic tomography would probably be the same.

2.7.6 Discussion

At the present time, to the authors' knowledge, no ultrasonic equipment is used in clinical situations for noninvasive thermometry during hyperthermia treatments. For historical reasons, mainly the accumulated experience and the availability of equip-

ment for biomedical applications, much effort has been devoted to both theoretical and experimental aspects of the modality. Despite encouraging results at the phantom experimentation or numerical simulation level, extension to the clinical level has not (yet) been successfully achieved. Many groups that have invested great effort in classical extensions of ultrasonic imaging techniques to noninvasive thermometry have now abandoned this direction.

Although very attractive with respect to their spatial resolution capabilities resulting from their small wavelength, ultrasonic waves do not appear suitable for detecting very small changes due to temperature elevation. In practice too many other parasitic factors, such as patient movements, shadow the desired signal.

While classical approaches appear difficult to use, it may be that the door is still open for thermo-induced acoustic imaging; a definite decision on its possible usefulness awaits adequate equipment.

2.8 Discussion, Synthesis, and Prospects

Noninvasive control is essential if hyperthermia is to realize its full potential as a therapy for malignant disease. This chapter has described techniques for noninvasive temperature monitoring and has presented results obtained by researchers with tissue-simulating phantoms or the human body. Considering the diversity of the modalities employed, and the wide variety of experimental setups used to test the apparatus, a quantitative comparison between these techniques is not, at present, feasible.

One major difficulty in comparing techniques is that they are all at different stages of development. Microwave radiometry has been used for many years in human body temperature measurement. X-ray CT, NMR, and ultrasound echography are well developed imaging techniques only recently considered for application in temperature monitoring. In contrast, electrical impedance tomography, active microwave imaging, and other ultrasound methods are relatively recent modalities at an early stage of development.

Another important consideration which we have attempted to address in this chapter is the extent to which each of the modalities measures temperature. All of the modalities considered, including microwave radiometry, are influenced to a greater or lesser degree by other factors, such as tissue composition and physiological changes, for example an increase in local blood flow, which may or may not vary during a hyperthermia treatment. In addition, temperature coefficients of each modality depend on tissue type. Unfortunately, few data exist on these factors and as such it is difficult to compare this feature of the different modalities. This is one area in which more investigation is needed.

A major conclusion which emerges from the research presented is the need for *imaging* of temperature distributions. Most of the techniques considered have an imaging capability and it is on the basis of temperature imaging that techniques can be compared. We consider the techniques in terms of temperature resolution, spatial resolution, measurement time, cost, and any other special features or limitations which pertain to specific modalities.

Microwave radiometry is so far the only noninvasive temperature monitoring technique to have been incorporated into a commercial hyperthermia system. To a certain extent, radiometry can act as a benchmark against which the other techniques can be evaluated. Microwave radiometry has proved useful for hyperthermia control in certain situations, notably in a semiquantitative control of superficial hyperthermia, especially where the radiometer probe can be incorporated into the hyperthermia applicator. Systems are relatively low cost in the single probe format, although proposed aperature synthesis radiometers may lead to large increases in cost. The major disadvantage with this technique is the limited depth of view, effectively limiting control to superficial tissue. Due to the complex interaction of microwaves with tissue, difficulties in interpreting results also occur. Other disadvantages include low data acquisition rates and interference from background electromagnetic noise which means that, for best results, it is often necessary to operate systems within screened chambers.

Of the techniques more recently considered for temperature monitoring, X-ray CT looks like being the first to produce a viable system for hyperthermia control. Its main advantage is that refined equipment already exists and image quality is good. Although the sensitivity of CT numbers to temperature is low, this is compensated for by good contrast resolution to give an adequate temperature resolution. The main concern with such a system is the ionizing radiation hazard. It is not clear whether safety standards will allow a frequent enough measurement of temperature distribution to allow adequate control of treatments. Other problems include possible movement artifacts both in the images and in heating procedure. The cost of an X-ray scanner is high but justifiable if the technique proves to be effective. However, there remains little scope for further improvement of the apparatus and this technique is likely to be superseded by others.

Another technique which benefits from refined equipment and excellent image quality is NMR. Although NMR does not suffer from ionizing radiation hazard, other factors do limit its applicability to temperature monitoring. Data acquisition time is long compared with X-ray CT, and the large magnet may lead to incompatibility with many heating systems. The present cost of an NMR facility is probably too high for a dedicated hyperthermia control system. Finally, the question of irreversibility of temperature response of NMR parameters has to be resolved.

Electrical impedance tomography (EIT) emerges as perhaps the most promising candidate for a routine hyperthermia control system. The technique's major advantage lies in its low cost, orders of magnitude lower than that of X-ray CT or NMR. EIT has only a moderate image quality when compared to these modalities but could give a medium resolution map of temperature distribution for a cost similar to a single probe radiometer. Other advantages include simplicity of use, very rapid data acquisition, good temperature resolution, and good compatibility with hyperthermia equipment. EIT is at a relatively early stage of development. However, with the rapid progress being made in data acquisition hardware and processing software, it seems likely that further research will soon produce a viable system which will allow improved hyperthermia control.

Active microwave imaging is among the least developed of all the techniques considered in this chapter but shows some promising features. Theoretical and experimental studies show a very favorable temperature resolution/rapidity trade-off when compared to microwave radiometry. Temperature resolution and data acquisition rates also compare well with the other modalities. The technique has a moderate image quality at a moderate cost. Compatibility with heating equipment may be a problem due to the need for a contacting bolus, but some configurations appear favorable. It is clear that this technique requires more development of equipment and especially of reconstruction algorithms.

Conventional ultrasound echography appears to be inappropriate for use in hyperthermia monitoring, due to a high sensitivity to patient movement amongst other factors. Research into this technique has been largely abandoned.

Finally, the following conclusions could be drawn:

1. The credibility of using hyperthermia for deep treatments depends on some kind of noninvasive

control. Without such control, it will be very difficult to decide whether poor results are to be explained by the inefficacy of hyperthermia or by the inadequacy of heating equipment.

2. Dielectric properties appear to be a sensitive means of visualizing deep thermal gradients. Furthermore, significant improvements can be reasonably expected from new emerging imaging techniques such as electrical impedance and microwave tomography. Nevertheless, whatever the selected technique, there is still a long way to go before it will be of clinical interest.

3. Between the present situation, where almost nothing is known except at a few points in the heated volume, and the ideal situation, where the temperature could be determined with 0.1 °C accuracy and 1 mm spatial resolution, there is room for modest but realistic and useful noninvasive thermal control equipment. From this point of view, the imaging techniques which have been reviewed in this chapter should be considered very interesting means of investigation despite the fact that they constitute poor thermometers.

4. The future will belong to those countries or companies which decide to invest in and support the long-term effort required to achieve clinical efficiency, overcoming the disappointment with respect to results obtained by means of hyperthermia over more than 10 years, as well as the unfavorable preconceived ideas on any modality that does not provide its ultimate level of performance immediately.

References

1. Adams MF, Anderson AP (1982) Synthetic aperture tomographic imaging for microwave diagnostics. Proc IEE 129 (2):83–88

2. Aida S, Iwama N, Ogura I (1985) Fundamental experiment of temperature dependence of ultrasound parameters. In: Egawa S (ed) Progress in hyperthermic oncology. Proc 2nd Annual Meeting Japan Soc Hyperthermic Onc 7–9 Nov, 1985, Shinshara Publishers Inc, pp 234–235

3. Amasha HM, Anderson AP, Conway J, Barber DC (1988) Quanitative assessment of impedance tomography for temperature measurements in microwave hyperthermia. Clin Phys Physiol Meas 9 [Suppl A]:49–53

4. Arcangeli G, Lombardini P, Lovisolo G, Marsiglia G, Piatelli M (1984) Focusing of 915 MHz electromagnetic power on deep human tissues: a mathematical model study. IEE Trans Bio Med Eng 31(1):47–52

5. Barber DC, Brown BH, Freeston IL (1983) Imaging spatial distributions of resistivity using applied potential tomography. Electronic Lett (19)22:933–934

6. Barber DC, Seager AD (1987) Fast reconstruction of resistance images. Clin Phys Physiol Meas 8 (Suppl)A: 47–54

7. Bardati F, Solimini D (1983) Radiometric sensing of biological layered media. Radio Sci 18 (6):1393–1401

8. Bardati F, Mongiaro M, Solimini D (1986) Synthetic array for radiometric retrieval of fields in tissues. IEEE Trans Microwave Theory Tech 34 (5)

9. Barret AH, Myers PC (1975) Subcutaneous temperatures: a method of non-invasive sensing. Science 190 (4215):669–671

10. Benjapolakul W, Shiina T, Kageyama Y, Saito M (1985) Non-invasive temperature measurement by ultrasound in hyperthermia. In: Egawas S (ed) Progress in hyperthermic oncology. Proc 2nd Annual Meeting Japan Soc Hyperthermic Onc, 7–9 Nov, 1985, Shinohara Publishers Inc, pp 232–233

11. Bentzen SM (1983) Quantitative computed tomography. Ph D Thesis, University of Aarhus

12. Bentzen SM, Overgaard J, Jorgensen J (1984) Isotherm mapping in hyperthermia using subtraction X-ray computed tomography. Radiother Oncol 2:255–260

13. Bolomey JC (1982) La méthode diffusion modulée: une aproche au relevé des cartes de champs microondes en temps réel. L'onde Electrique 62(5):73–78

14. Bolomey JC, Peronnet G, Pichot C, Jofre L, Gautherie M, Guerquin-Kern JL (1984) L'imagerie microonde active en génie biomédical. In: Lewiner J (ed) L'imagerie du corps humain. Physique, les Ulis, France

15. Bordure G, Delauzin JP, Dubois JB, Hay M (1985) Application et contrôle atraumatique de l'hyperthermie par microondes pour le traitement de tumeurs superficielles. J Biophys Biomech 9:37–39

16. Bowen T (1982) Radiation induced thermoacoustic imaging. US Patent, PCT, US 82/00495

17. Brooks LA, Mitchell LG, O'Connor CM, DiChiro G (1981) On the relationship between computed tomography numbers and specific gravity. Phys Med Biol 26 (1):141–147

18. Brown BH, Barber DC, Seager AD (1985) Applied potential tomography: possible clinical applications. Clin Phys Physiol Meas 6:109–121

19. Bruggmoser G, Hinkelbein W (1986) The applicability of microwave thermography for deep-seated volumes. In: Brugmoser G et al. (eds) Recent results in cancer research, vol 101. Springer Berlin Heidelberg New York, pp 88–98

20. Burdette EC, Cain FL, Seals J (1980) In-vivo probe measurement technique for determining dielectric properties at VHF through microwave frequencies. IEEE Trans Microwave Theory Tech 28(4):14–427

21. Burdette EC (1983) Influence of blood flow on tissue electrical properties: examination of regional blood flow in the dog kidney by a new probe method. Ph D Thesis, Emory University School of Medicine, Georgia

22. Bydder GM, Kreel L (1979) The temperature dependence of computed tomography attenuation values. J Comput Assist Tomogr 4:506–510

23. Caspers F, Conway J (1982) Measurement of power density in a lossy material by means of electromagnetically induced acoustic signals for non-invasive determination of spatial thermal absorption in connection with pulsed hyperthermia. In: 12th-European Microwave Conference Helsinki, 13–17 Sept, 1982, Microwave Exhibitions and Publishers Ltd, pp 565–568

24. Cetas C (1984) Will thermometric tomography become practical for hyperthermia treatment monitoring? Cancer Res (Suppl)44:4805s–4808s

25. Cheung AY, Golding WM, Samaras GM (1981) Direct contact applicator for microwave hyperthermia. J Microwave Power 16(2):151−159

26. Chivé M, Plancot M, Leroy Y, Giaux G, Prévost B (1982) Microwave and radiofrequency hyperthermia monitored by microwave thermography. In: 12th European Microwave Conference Helsinki, 13−17 Sept, 1982, Microwave Exhibitions and Publishers Ltd, pp 547−552

27. Chivé M (1985) Technical aspects of microwave hyperthermia controlled by microwave radiometry. Odam-Brücker Medical Report 85(1):30−34

28. Christensen DA (1982) Current techniques in non-invasive thermometry. In: Nussbaum GH (ed) Physical aspects of hyperthermia, médical physics monograph no 8. Amercian Institute of Physics, New York, pp 266−279

29. Coldefy H (1986) Contrôle non-invasif de l'hyperthermie par imagerie microonde active. Etude préliminaire sur fantôme homogène. Ph D Thesis Université de Paris-Sud

30. Conway J, Hawley MS, Seagar AD, Brown BH, Barber DC (1985) Applied potential tomography (APT) for non-invasive thermal imaging during hyperthermia treatment. Electronic Lett 21:836−838

31. Conway J (1987) Electrical impedance tomography for thermal monitoring of hyperthermia treatment: an assessment using in vitro and in vivo measurements. Clin Phys Physiol Meas 8, [Suppl A]:141−146

32. David BJ, Lele PP (1985) An acoustic phase shift technique for the non-invasive measurement of temperature changes in tissues. Proc. IEEE 1985, Ultrasonics Symp San Francisco, CA

33. de Lateur BJ, Lehman JF, Stronebridge JB, Warren CG, Guy AW (1970) Muscle heating in human subjects with 915 MHz microwave contact applicator. Arch Phys Med 51:147

34. de Talhouët H (1986) Contribution à l'amélioration de la résolution en imagerie microonde monochromatique. Ph D Thesis, Université de Paris-Sud

35. Devaney AJ (1982) A filtered back-propagation algorithm for diffraction tomography. Ultrasonic Imaging 4:336−350

36. Dickinson RJ, Hall AS, Hind AJ, Young IR (1986) Measurement of changes in tissue temperature using MR images. J Comput Assist Tomogr 10(3):468−472

37. Do-Huu JP, Mayeux C, Micheron F (1985) In-vivo effect of human tissue heating on NMR images. VIIth Meeting of ESHO, Paris

38. Duchêne B, Tabbara W (1985) Tomographie ultrasonore par diffraction. Rev Phys Appl 20:299−304

39. Edrich J, Jobe WE (1982) Imaging microwave thermography. Temperature (Am Inst Phys) 5:1379−1380

40. Edrich J, Hardee PC (1974) Thermography at millimetre wavelengths. Proc IEEE 62 (10):1391

41. Enel L, Leroy Y, Van de Velde JC, Mamouni A (1984) Improved recognition of thermal structures by microwave radiometry. Electronic Lett 20(7):293−294

42. Ermert H, Dohlus M (1986) Microwave-diffraction-tomography of cylindrical objects using 3-dimensional wave-fields. NTZ Archiv 8(5):111−117

43. Eyüboglu BM, Brown BH, Barber DC, Seager AD (1987) Localisation of cardiac related impedance changes in the thorax. Clin Phys Physiol Meas 8 [Suppl A]:167−173

44. Fallone BG, Moran PR, Podgorsak EB (1982) Non-invasive thermometry with a clinical X-ray CT scanner. Med Phys 9(5):715−721

45. Gernero LH (1987) Reconstruction d'images tomographiques à partir d'un ensemble limité de projections. Ph D Thesis, University of Paris, Paris

46. Godgaonkar DK, Ghandi OP, Hagmann MJ (1983) Estimation of complex permittivities of three-dimensional biological bodies. IEEE Trans Microwave Theory Tech 31 (6):442−446

47. Grant EH (1984) Dielectric properties of normal and malignant tissues. In: Colloquium on Electromagnetic techniques for the detection and treatment of malignant disease. IEE Digest (36):1

48. Griffiths H, Ahmed A (1987) Applied potential tomography for non-invasive temperature mapping in hyperthermia. Clin Phys Physiol 8 [Suppl A]:147−153

49. Guerquin-Kern JL (1980) Hyperthermie locale par microondes en thérapeutique cancérologique. Thèse 3ème Cycle, University of Strasbourg, Strasbourg

50. Guerquin-Kern JL, Gautherie M, Peronnet G, Jofre L, Bolomey JC (1985) Active microwave tomographic imaging of isolated perfused animal organs. Bioelectromagnetics 6:145−146

51. Guo TC, Guo WW, Larsen LE (1984) A local field study of a water-immersed microwave antenna array for medical imagery and therapy. IEEE Trans Microwave Theory Tech 32(8):844−860

52. Guo TC, Guo WW, Larsen LE (1986) Recent developments in microwave medical imagery. Phase and amplitude conjugations and the inverse scattering theorem. In: Larsen LE, Jacobi J (eds) Medical applications of microwave imaging. IEEE P, New York, pp 167−183

53. Guo TC, Guo WW (1987) Physics of image formation by microwave scattering. Medical Imaging, SPIE Proc 767, (2):816−819

54. Guy AW (1971) Analyses of electromagnetic fields induced in biological tissues by thermographic studies on equivalent phantom models. IEEE Trans Microwave Theory Tech 19(2):205−213

55. Hagmann M (1981) Application of moment methods to electromagnetic biological imaging. Proc IEEE MTT-S Symposium, Los Angeles, 15−19 June 1981, p 482

56. Hand JW (1984) Thermometry in hyperthermia. In: Overgaard J (ed) Hyperthermia oncology, vol 2. Taylor and Francis, London, pp 299−308

57. Haney MJ, O'Brien WD (1982) Ultrasonic tomography for differential thermography. In: Ash EA, Hills CR (eds) Acoustic imaging. Plenum, New York, pp 589−597

58. Harrington RF (1961) Time harmonic electromagnetic fields. McGraw-Hill, New York

59. Harris ND, Suggett AJ, Barber DC, Brown BH (1987) Application of applied potential tomography (APT) in respiratory medicine. Clin Phys Physiol Meas 8 [Suppl A]:155−165

60. Haslam NC, Gillespie AR, Haslam CGT (1984) Aperture synthesis thermography: a new approach to passive microwave temperature measurement in the body. IEEE Trans Microwave Theory Tech 32(8):829−835

61. Hawley MS (1986) Microwave radiometric thermometry in layered tissue structures. PhD Thesis, University of Sheffield, Sheffield

62. Hawley MS, Conway J, Anderson AP, Cudd PA (1988) The influence of tissue layering on microwave thermographic measurements. Int J Hyperthermia 4(4):427−435

63. Hay M, Dubois JB, Bordure G (1987) Applicateurs à géométrie variable et contrôle atraumatique de l'hyperthermie par microondes de 2450 MHz dans le traitement des tumeurs superficielles. Innov Tech Biol Med 8 (3): 294−305

64. Hessary MK, Chen KM (1984) EM local heating with HF electric fields. IEEE Trans Microwave Theory Tech 32 (6):569−576

65. Hirai S, Nikawa Y, Okada F, Kikuchi M, Mori S (1987) Dual waveguide applicator with temperature measurement in EM hyperthermia. IEEE 9th Annual Conf IEEE-EMB Soc, pp 1308–1309

66. Jacobi JH, Larsen LE, Hast CT (1979) Water-immersed microwave antennas and their application to microwave interrogation of biological targets. IEEE Trans Microwave Theory Tech 27(1):70–78

67. Jacobi JH, Larsen LE (1980) Microwave time-delay spectroscopic imagery of isolated canine kidney. Med Phys 7 (1):1–7

68. Jofre L, Reyes E, Ferrando M, Elias A, Romeu J, Baquero M (1986) A cylindrical system for quasi-real time microwave tomography. 16th European Microwave Conference, Dublin, pp 599–604

69. Kaveh M, Soumekh M, Greenleaf JF (1984) Signal processing for diffraction tomography. IEEE Trans SUU-31 (4):230–239

70. Knüttel B, Juretschke HP (1986) Temperature measurement by nuclear magnetic resonance and its possible use as a means of in-vivo non-invasive temperature measurement and for hyperthermia treatment assessment. Recent Results Cancer Res 101:109–118

71. Krug J, Edenhofer P (1985) Microwave acoustic imaging for medical applications. 17th European Microwave Conference, Paris, 9–13 Sept 1985, pp 655–660

72. Krug J (1987) Private communication

73. Landau LD, Lifshitz EM (1960) Electrodynamics of continuous media. Pergamon, New York

74. Larsen LE, Jacobi JH (1979) Microwave scattering parameter imagery of an isolated canine kidney. Med Phys 6 (5):394–403

75. Larsen LE, Jacobi JH (1982) Microwaves offer promise as imaging modality. Diagn Imag Clin Med 11:44–47

76. Lewa J, Majewska Z (1980) Temperature relationship of proton spin-lattice relaxation time T1 in biological tissues. Bull Cancer 67(5):525–530

77. Ludeke KM, Koehler J, Kanzenbach J (1979) A radiation balanced microwave thermograph for medical applications. Acta Electronica 22(1):65–69

78. Mamouni A, Leroy Y, Van de Velde JC, Bellardi L (1983) Introduction to correlation microwave thermography. J Microwave Power 18(3):285–293

79. Man (1974) ICRP 23, Pergamon Press, Oxford

80. Milligan AJ, Couran PB, Ropar MA, McCulloch HA, Ahuja RK, Dobelbower RR (1983) Predictions of blood flow from thermal clearance during regional hyperthermia. Int J Radiat Oncol Biol Phys 9:1335–1343

81. Mizushina S (1987) Automedica 8 (4) (special issue on noninvasive temperature measurement)

82. Mueller RK, Kaveh M, Wade G (1979) Reconstruction tomography and application to ultrasonics. Proc IEEE, 67 (4):567–587

83. Nasoni RL, Bowen T, Conner WG, Sholes RR (1979) In-vivo temperature dependence of ultrasound speed in tissue and its applications to non-invasive temperature monitoring. Ultrason Imaging 1(1):34–414

84. Newman WH, Dittwar A, Delhomme G, Delannoy J (1986) Tumor perfusion during microwave hyperthermia: preliminary measurements. Proc IEEE 8th Ann Conf EMB Soc pp 1503–1506

85. Olsen RG, Lin JC (1981) Microwave pulse induced resonances in spherical head models. IEEE Trans Microwave Theory Tech 29(10):1114–1117

86. Olsen RG, Lin JC (1983) Acoustical imaging of a model of human hand using pulsed microwave irradiation. Bioelectromagnetics 4:397–400

87. Pichot C, Jofre L, Peronnet G, Bolomey JC (1985) Active microwave imaging of inhomogeneous bodies. IEEE Trans AP 33(4):416–425

88. Plancot M, Prévost B, Chivé M, Fabre JJ, Ledel I, Giaux G (1987) A new method for thermal dosimetry in microwave hyperthermia using microwave radiometry for temperature control. Int J Hyperthermia 3(1):9–19

89. Rajagopalan B, Greenleaf JF, Thomas PJ, Johnson JA, Bahn RC (1979) Variation of acoustic speed with temperature in various excised human tissues studied by ultrasound computerized tomography. In: Linzer M (ed) Ultrasound tissue characterization. US Gov Printing Office, NBS Special Publication, Washington DC, 525, pp 227–233

90. Rangayyan RM (1986) Computed tomography techniques and algorithms: a tutorial. Innov Tech Biol Med 7(6): 746–762

91. Robert J, Marchal C, Escanye JM, Thouvenot P, Gaulard ML, Tosser A (1981) Ultrasound velocimetry for hyperthermia control. Prog Clin Biol Res 107:555

92. Robillard M (1981) Contribution à l'étude des sondes et à la reconnaissance d'objet thermique par la thermographie microonde. Thèse 3ème Cycle, Université de Lille, Lille

93. Roubine E, Bolomey JC (1977) Antennes. Masson, Paris

94. Seagar AD, Brown BH (1987) Limitation in hardware design in impedance imaging. Clin Phys Physiol Meas 8 [Suppl A]:85–90

95. Seagar AD, Barber DC, Brown BH (1987) Theoretical limits to sensitivity and resolution in impedance imaging. Clin Phys Physiol Meas 8 [Suppl A]:13–31

96. Slaney M, Kak AC, Larsen LE (1984) Limitations of imaging with first-order diffraction tomography. IEEE Trans Microwave Theory Tech 32(8):860–874

97. Tarassenko L, Rolfe P (1984) Imaging spatial distributions of resistivity – an alternative approach. Electronic Lett 20 (14):574–575

98. Van Hippel (1955) Dielectric materials and applications. Wiley, New York

99. Yorkey TJ, Webster JG (1987) A comparison of impedance tomographic reconstruction algorithms. Clin Phys Physiol Meas 8 [Suppl A]:55–62

100. Zamenhof RG, Sternick ES, Curran B (1983) Comments on "non-invasive thermometry with a clinical X-ray CT scanner". Med Phys 10(3):374

101. Zamenhoff RG, Sternick ES, Curran BM (1981) Non-invasive temperature mapping by computerized tomography. Int J Radiat Oncol Biol Phys 7:1235

102. Zheng E, Shao S, Webster JG (1984) Impedance of skeletal muscle from 1 Hz to 1 MHz. IEEE Trans Biomed Eng 31 (6):477–481

103. Parker DL, Smith V, Sheldon P, Crooks LE, Fussell L (1983) Temperature distribution measurements in two-dimensional NMR imaging. Med Phys 10 (3):321–325

3 Use of Microwave Radiometry for Hyperthermia Monitoring and as a Basis for Thermal Dosimetry

M. Chive

3.1 Introduction

Microwave radiometry originated in Dicke's work on atmospheric absorption (1946–1947) [1]. The technique he described has been and still is used in radio astronomy, but it can also be applied to other domains, such as:

1. The study of the propagation of electromagnetic waves in the atmosphere
2. The teledetection of resources on earth
3. The recovery af avalanche victims
4. Meteorology, glaciology, and oceanography
5. Missile monitoring
6. More recently, medical applications

The application of microwave radiometry to medicine and biology stems from the fact that plotting a thermal subcutaneous map of living tissues is of permanent concern and is solved only with difficulty by means of invasive methods of thermocouple implantation within tissues.

The microwave radiometry technique, in which trauma is totally absent, can yield information in three types of study:

1. Physiological studies on the mechanism of thermoregulation and metabolism
2. Diagnostic studies involving cancerous or rheumatic diseases
3. Temperature control and monitoring of hyperthermia sessions

The medical applications of microwave radiometry date back to 1974–75 in the United States [2], and research into such applications commenced in France in 1978–79 [3]. This research led to the production of the TMO and HYLCAR systems by the firm ODAM (Wissembourg, France), which are now in clinical use. This chapter will discuss the physical principles of thermal radiation of electromagnetic origin, the principle and structure of radiometers, and finally dosimetry in hyperthermia based on radiometric temperature measurements.

3.2 Measurement of Thermal Radiation

3.2.1 Physical Principles: The Black Body Radiation

A given body whose temperature is raised to 0K emits a spontaneous electromagnetic radiation of thermal origin. In the case of the black body, perfectly absorbing and nonreflecting, spectrum brightness $B(f, T)$ (i.e., energy radiated per unit of apparent surface and per unit of solid angle), at a frequency f and for a bandwidth of 1 Hz, is expressed by Planck's law:

$$B(f, T) = \frac{2hf^3}{C^2\{\exp[-(hf/kT)] - 1\}} \qquad (3.1)$$

where h = Planck's constant, k = Boltzmann's constant, T = temperature (K), f = frequency (Hz), and C = speed of light.

Figure 3.1 represents $B(f, T)$ for a black body whose temperature has been raised to 310K (or 37°C), a

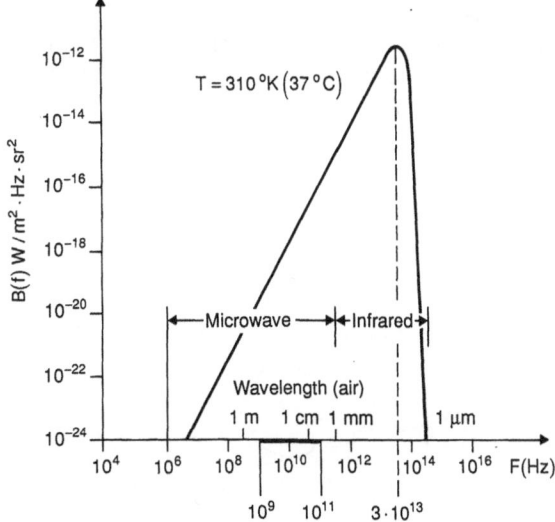

Fig. 3.1. Spectrum brightness versus frequency for a black body whose temperature is 310K

temperature close to that of living tissues. This curve shows a peak in the infrared area. In the domain of Hertz waves (f < 300 GHz) when $hf \ll kT$, Planck's law can be simplified and leads to the Rayleigh Jeans relation:

$$B(f, T) = 2kTf^2/C^2 \ . \tag{3.2}$$

This equation shows that the energy radiated by the emissive black body is directly proportional to temperature. The temperature of an emissive body can thus be determined by a measurement of the electromagnetic power radiated in a given bandwidth. This measurement is achieved by radiometric systems which use an antenna as an electromagnetic power captor in the microwave region.

3.2.2 Thermal Power Collected by an Antenna: Nyquist's Formula

Let us consider an antenna pointing to a black body whose temperature T is uniform and which is supposed to be in thermodynamic equilibrium as compared to the surrounding medium (Fig. 3.2). This antenna is defined by its gain, its radiation diagram, and its bandwidth. If the aperture angle θ_0 or separating power of the antenna is narrower than angle θ at which the black body can be seen, it is to be noted that, when Rayleigh Jeans' law can be applied, the power P collected by the antenna in a bandwidth Δf is given by Nyquist's formula:

$$P = k \cdot T \cdot \Delta f \tag{3.3}$$

This power is independent of the frequency and the dimensions of the antenna.

Note: if the black body is seen at an angle $\theta < \theta_0$ then the power collected by the antenna is expressed by the following relation:

$$P \# k \cdot T \cdot \Delta f \left(\frac{\theta}{\theta_0}\right)^2 \ . \tag{3.4}$$

3.2.3 Thickness of Medium and Radiometric Measurement

A given body is characterized by its dielectric constants ε' and ε'' (or by $\varepsilon^* = \varepsilon' - j\varepsilon''$) which depend on frequency and temperature in general. Starting from these constants the attenuation α of the electromagnetic waves absorbed by the medium can be calculated. This attenuation is identical to that of electromagnetic waves of thermal origin emitted by a body at temperature T. If all the particles of the medium (supposed to be nonreflecting) take part in the thermal emission, the contribution of each of them to the signal emitted outward can be very different. In fact, the electromagnetic radiation emanating from a point situated at a distance z from the surface is attenuated according to $\exp(-\alpha z)$ before reaching free space. Thus only the medium layer whose thickness is 2 or 3 times that of penetration depth ($\delta = 1/\alpha$) contributes to the signal collected outside.

Let us point out here the identity which can be noted between the absorption phenomena of an electromagnetic wave and those of a thermal emission for the same emissive material and same frequency or bandwidth: in both cases the same volume is concerned by the two phenomena.

So in the TEM mode – i.e., for a plane wave in free propagation – a layer of the homogeneous medium whose thickness Δz is raised to a temperature $T + \Delta T$ (Fig. 3.3) will radiate outward such an electromagnetic power that the apparent temperature will be:

$$T_A = T + \Delta T_A \tag{3.5}$$

with $\Delta T_A = \Delta T \exp(-\alpha z) [1 - \exp(-\alpha \Delta z]$.

The medium volume undergoing radiometric measurement depends on attenuation α and thus on the penetration depth at the considered frequency.

For biological media (muscular tissue, skin, fat, etc.) in the infrared range δ is very slight (< 1 mm) while in the microwave range living tissues, which are bodies

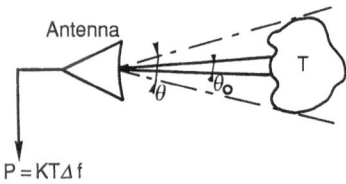

Fig. 3.2. Radiated power collected by an antenna from a black body raised at a temperature T

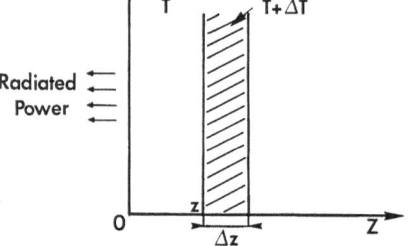

Fig. 3.3. Calculation of the apparent temperature deduced from the thermal radiated power

of much lower absorption, offer much higher penetration depths.

If the medium dielectric constants are ε' and ε'', the expression of the attenuation α that a plane wave undergoes when propagating through the medium is as follows:

$$\alpha_E = \frac{2\pi f}{C} \sqrt{\tfrac{1}{2}(-\varepsilon' + \sqrt{\varepsilon'^2 + \varepsilon''^2})} \quad \text{Neper/m} \quad (3.6)$$

This is the attenuation of the electric field. Power attenuation α is:

$\alpha = 2\alpha_E$ used to define the penetration depth in power

$\delta = 1/\alpha = 1/2\alpha_E$ in meters

Figure 3.4 shows the frequency evolution of δ for the two most common types of tissue: muscle which contains a high percentage of water and fat [4, 5]. It should be stressed that these values concern a plane wave TEM and can only be applied rigorously to the detection of a normal beam at the air-medium diopter under investigation when the antenna is removed from the medium.

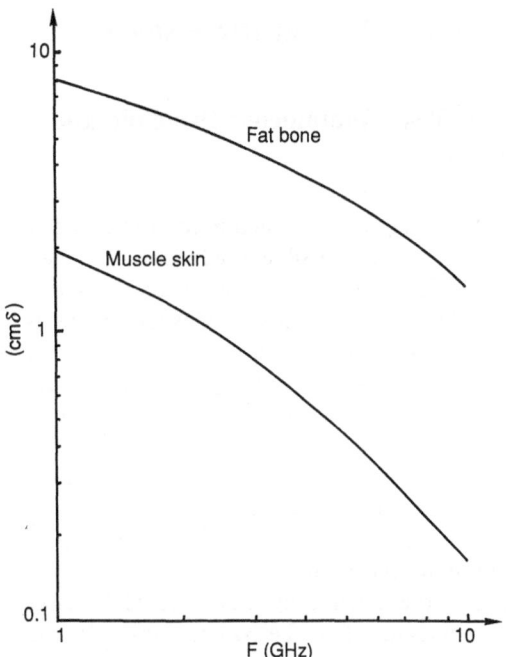

Fig. 3.4. Frequency evolution of the penetration depth (in power) for muscle or skin and fat or bone in the case of a plane wave propagation

3.2.4 Reflection Effects at the Air-Medium Interface

As a matter of fact the energy transfer from the emissive body to the reception antenna is not thoroughly achieved as in the case of the black body. What must be taken into account are the reflection effects at the air-medium interface, which are characterized by the power reflection coefficient ϱ; this expresses the percentage of power that does not radiate out of the diopter and is sent back inward.

Nyquist's relation must then be written as follows:

$$P = (1-\varrho)\, k \cdot T \cdot \Delta f \quad (3.7)$$

where $(1-\varrho)$ is the emissivity.

ϱ is directly linked to the complex permittivity ε^* of the two media concerned:

$$\varrho = \left[\frac{\sqrt{\varepsilon_1^*} - \sqrt{\varepsilon_2^*}}{\sqrt{\varepsilon_1^*} + \sqrt{\varepsilon_2^*}}\right]^2 . \quad (3.8)$$

Index 1 is linked to the emission medium, index 2 to the reception medium.

Note. These reflection effects can be felt when the signal crosses two tissue layers of different nature (fat/skin, muscle/fat, etc.). Thus these effects take part in the thermal emission. Another consequence is that the antenna may collect signals that are being reflected by the body but emanate from other sources.

3.2.5 Characteristics of the Radiometric Method for Temperature Measurement

Choice of frequency. From 1 to 10 GHz considering the attenuation values α of the biological tissues, thus considering δ.

Sensitivity. The radiometric receiver must be able to detect temperature variations of 0.1 °C, which correspond to a sensitivity ΔP

$$\Delta P = k\,\Delta T \cdot \Delta f = 10^{-15}\,\text{W} \quad (3.9)$$

with $\Delta f = 1$ GHz, assuming a perfect adaptation to the medium (i.e., $\varrho = 0$).

In order to avoid unwanted signals the reception antenna will be put in contact with the medium and will be designed so as to have a reflection coefficient at most equal to 0.1 in a large bandwidth (1 GHz) around the chosen central frequency of the radiometer. This will free the device of air-medium interface problems.

3.3 Microwave Radiometric Systems

3.3.1 The Dicke Radiometer: Principle and Limitations

In 1946, Dicke suggested a structure for a radiometer (Fig. 3.5) which made possible a reduction of the effects of amplification chain gain fluctuations and permitted a fairly good sensitivity. This radiometer is characterized by a modulator fitted between the antenna and the receiver (amplifier + microwave detector). Channel 1 of the modulator is connected to the antenna, channel 2 to a well-matched reference charge (or to a noise source) raised to temperature T_R. The cyclic pulse generator whose ratio is $\frac{1}{2}$ monitors the synchronicity of the modulator and of the synchronous detection.

If we replace the antenna by a $50\,\Omega$ coaxial charge raised to temperature T, at the system output, the signal is proportional to the difference in temperature $T-T_R$. In fact:

$$P_S \sim G(T+T_B) - G(T_R+T_B) \qquad (3.10)$$

State 1 of the   State 2 of the
modulator   modulator

or also

$$P_S \sim G(T-T_R\ .) \qquad (3.11)$$

G and T_B are respectively the chain gain and the

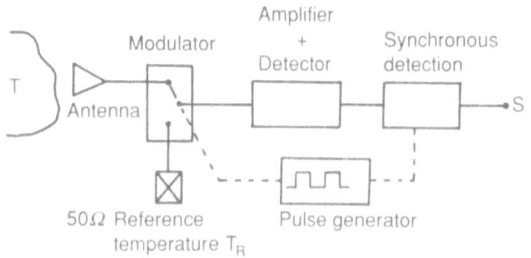

Fig. 3.5. Structure of the Dicke radiometer

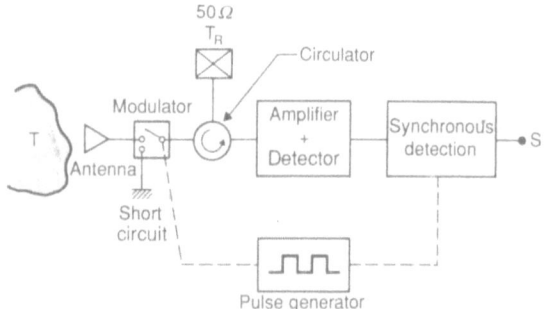

Fig. 3.6. Structure of the modified radiometer

receiver noise temperature. T_B can be calculated from the noise factor of the receiver (F in db):

$$T_B = T_0\ 10^{F/10} \qquad (3.12)$$

with T_0 = ambient temperature.

The switching frequency state 1, state 2 is rather high since G is assumed to be constant. Nevertheless, the variations of G entail variations of P_S, which is the output signal.

If the value of the output signal is assumed to be zero by varying T_R, $P_S \sim 0$ when $T_R = T$ according to Fig. 3.6. This depicts a Dicke radiometer to which a ferrite circulator has been fitted between the modulator and the amplifier. The circulator is closed by the reference noise source T_R. Moreover, channel 2 of the modulator is connected to a short circuit. If we assume the modulator and the circulator to be perfect (i.e., no insertion noise) the power collected at the amplifier input is the following.

When the modulator is in state 1:

$$P_1 = k\varDelta f\ [(1-\varrho)T + \varrho T_R]\ . \qquad (3.13)$$

When it is in state 2:

$$P_2 = k\varDelta f T_R\ . \qquad (3.14)$$

The cyclic ratio of control pulses being $1/2$ after amplification and detection, the output signal S is proportional to $P_2 - P_1$, i.e.,

$$S = K(1-\varrho)[T-T_R]\ . \qquad (3.15)$$

By adjusting T_R so as to obtain a zero detected power (zero method), T can be deducted whatever ϱ:

$$S = 0\quad \text{when}\quad T = T_R\ .$$

Then the output signal fluctuations become independent from the gain fluctuations:

$$\varDelta P_s \sim \varDelta G(T-T_R) + G(\varDelta T - \varDelta T_R)\ . \qquad (3.16)$$

At $T = T_R$ then $\varDelta P_S \sim G(\varDelta T - \varDelta T_R)\ .$

The theoretical sensitivity of the Dicke radiometer then reads:

$$\varDelta T = \frac{\sqrt{2}(T+T_B)}{\sqrt{\varDelta f t_c}} \qquad (3.17)$$

where t_c = time constant of the synchronous detection, $\varDelta f$ = receiver bandwidth, and T_B = chain noise temperature.

Yet associating the Dicke radiometer to an antenna or applicator does not necessarily lead to determination of the temperature of the emissive medium facing the antenna. Equation 3.11 can only be applied when the antenna is perfectly suited to the medium, which can never be thoroughly achieved since the antenna facing the medium has a reflection coefficient ϱ which must be taken into account. Equation 3.11 then becomes

$$P_S \sim G(1-\varrho)(T-T_R) \ . \tag{3.18}$$

The Dicke radiometer thus does not measure the medium temperature but an apparent temperature $T_{app} = (1-\varrho)T$, which depends on the medium emissivity $(1-\varrho)$.

3.3.2 The Modified Radiometer

3.3.2.1 Principle

In order to free the system from the reflection coefficient at the air-emissive medium interface, the research team of the Centre Hyperfréquences et Semiconducteurs (Lille, France) has developed and patented [6–8] a modified radiometer as depicted in Fig. 3.6. This modified radiometer yields measured temperature values that are closer to T than those obtained by the Dicke radiometer since the imperfections of the components are taken into account.

3.2.2.2 Zero Method Limitations: Assessment of Error Due to Component Imperfections

The cable link between the antenna and the modulator is a microwave component which brings about insertion losses. According to the concept of radiative transfers, each of the elements attenuates the signal it carries and also generates a thermal signal which depends on its temperature and its attenuation. D. D. Nguyen [9] calculated the error these elements bring about when the output signal being zero, T is equal to T_R.
If $T_R = T_0 + \Delta T_R$ with T_0 = the ambient temperature:

$$T = T_0 + \Delta T \tag{3.19}$$

t_1 = transmission coefficient of the modulator,
t_2 = transmission coefficient of the cable,
ϱ = reflection coefficient of the antenna-emissive medium.
We then obtain:

$$\Delta T_R \# \frac{t_2}{t_1}\left(\frac{1-\varrho}{1-\varrho t_2^2}\right) \Delta T \ . \tag{3.20}$$

The problem of the losses due to the cable link between the antenna and the modulator then proves important. In the X band (8–12 GHz) for instance, a 1 m long cable is known to attenuate the signal between 1.5 and 2.5 db (or t_2 between 0.7 and 0.56) whereas with the modulator the losses remain in the region of 0.3 db at most (or $t_1 = 0.93$). These insertion losses can be determined once and for all, but the error made by equating ΔT_R to ΔT still depends on the reflection coefficient ϱ. In order to minimize the error due to the reflection coefficient a cable can be inserted between the short circuit and input 2 of the modulator. Its length and characteristics are the same as those of the cable used in channel 1 between the antenna and the modulator. With such a design Nguyen demonstrated that the output signal could then read:

$$S \sim K(1-\varrho)[t_1 t_2 \Delta T_R - \Delta T]t_1 t_2 \ . \tag{3.21}$$

Thus the error committed on ΔT when applying the zero method (S = 0 at the output) stops depending on the reflection coefficient since we have:

$$\Delta T \# t_2 t_1 \Delta T_R \ . \tag{3.22}$$

This is an interesting result since it then suffices to assess once and for all the transmission of the modulator and of the cables and to take these values into account to adjust the synchronous detection. For this purpose a preliminary calibration of the radiometer is carried out by replacing the antenna by a 50 Ω coaxial charge thermostated at different temperatures in order to adjust the chain gain so that the calibration slope of the radiometer is equal to one: output signal variations $\Delta S = \Delta T$ adapted charge temperature variation.

3.3.2.3 The Pseudo-zero Method Applied to the Medical Radiometer

The zero method requires a "heavy" implementation, since by varying the thermal noise power of reference T_R a zero value must be obtained at the synchronous detection output. Yet in the medical domain the subcutaneous temperatures to be measured range from 32 °C to 38°–39 °C for diagnostic purposes at to 44°–45 °C in the case of hyperthermia. The reference temperature T_R is then set around 38 °C. In these conditions, the maximum variation $T-T_R$ to be measured will be about 6 °C. In this case the error possible when determining $T-T_R$ is all the more slight since the reflection coefficient ϱ is low.
As a consequence applicator antennas have been designed so as to offer a reflection coefficient with tissues that is less than 0.1 (or < – 10 db). Moreover a preliminary calibration of the radiometer frees the system from these problems: the gauging here is car-

ried out by putting the applicator in contact with a liquid emissive medium (water + physiological serum) that simulates the biological tissues whose temperature is made to vary. We then get a gauging curve $S_{rad} = f(T)$ and an electronic device that complements the radiometer and makes possible the immediate display of the temperature.

3.3.3 Description and Performances of the Radiometer Used for Medical Applications; Performances of the Applicator Antenna

3.3.3.1 Description and Performances

The radiometers used for medical purposes developed by the Lille research team and now produced by the firm ODAM (Wissembourg, France) work according to the pseudo-zero method in the following bandwidths: 0.8 – 2 GHz, 2 – 4 GHz, and 8 – 10 GHz. They are either of the direct amplification kind (in the 8 – 10 GHz band) or of the superheterodyne kind (in the other frequency ranges) (Fig. 3.7). All are equipped with a pin diode modulator, a wide bandwidth ferrite circulator with one channel closed on to a 50 Ω coaxial charge thermostated so as to act as the reference thermal noice source, and a low now high gain amplifier (noise factor = 3 db; gain ⩾ 35 db).

The amplified signal is carried out to a balanced mixer in the case of the superheterodyne receiver. The

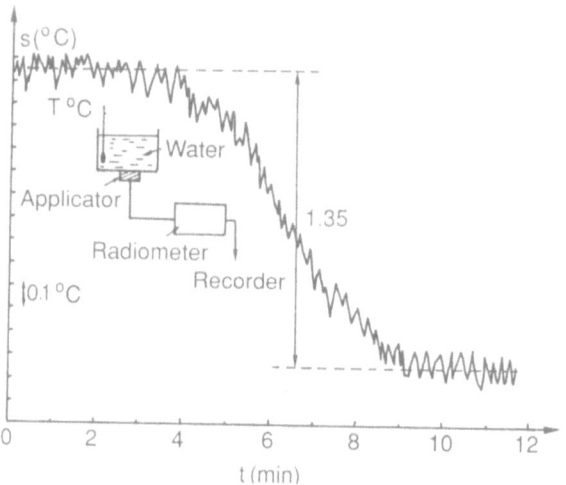

Fig. 3.8. Experimental sensitivity of the 1-GHz radiometer: recording of the radiometric temperature versus time

output signal at the intermediate frequency (IF) is first amplified, then detected before being processed by the synchronous detection.

Altogether, the chain must have a 70 db minimum gain, a 1-GHz bandwidth, and a noise factor inferior to 6 db in order to obtain a sensitivity of around 0.1 °C for a 2 s measurement time constant. The system output signal is carried out to a display unit that allows immediate read out of the temperature.

Figure 3.8 shows an example of the recorder radiometric signal versus time obtained by using a radiometer set to around 1 GHz. The applicator, insulated by a Mylar sheet, is placed on a thermostated bath of water whose temperature has been raised by 1 °C after a certain time: here the experimental sensitivity is about 0.1 °C.

The calibration of radiometers is achieved by measuring the temperature of a mixture of water and physiological serum that simulates such tissues as skin or muscle; consequently the radiometer output signal depends on the temperature of the liquid medium.

3.3.3.2 The Applicator

One of the essential elements of the radiometer is the applicator, which must be almost perfectly suited to the tissues it is in contact with. At the moment the applicators used are of the microstrip-microslot type [10, 11]. They are lighter, more efficient, and more cheaply produced than waveguides filled with lossy solid dielectric material with high permittivity. The applicator we have developed is a classical structure which consists of a feeding microstrip line and a slot

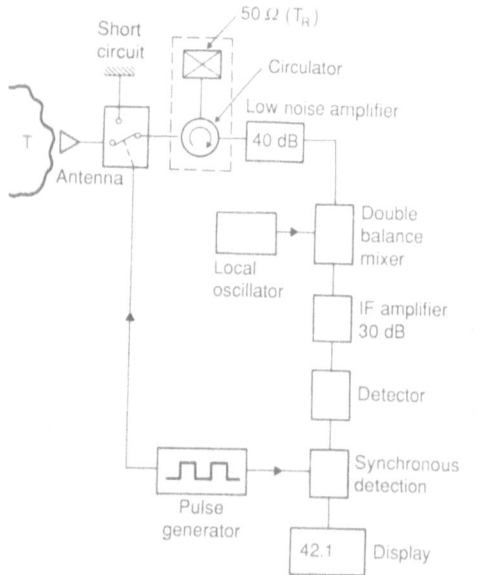

Fig. 3.7. Block diagram of the modified radiometer used for medical applications

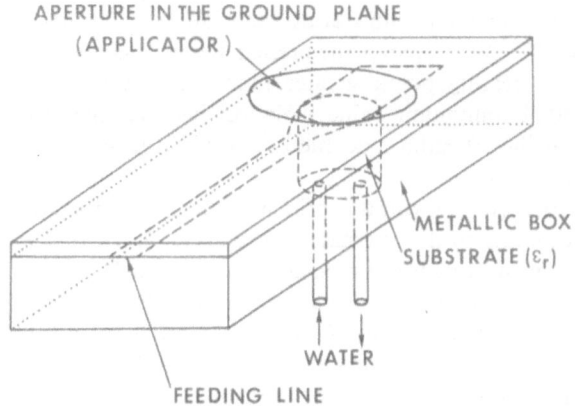

APERTURE IN THE GROUND PLANE
(APPLICATOR)

METALLIC BOX
SUBSTRATE (ε_r)

WATER

FEEDING LINE

Fig. 3.9. Schematic of the microstrip-microslot applicator with central cooling system

line opened in the ground plane to constitute the applicator aperture in contact with tissue. We have shown that the coupling of several slot line resonators opened in the ground plane of a microstrip line allows a good matching in wide bandwidth to be obtained when the applicator is in contact with tissue [12, 13]. From these early results we have designed the new type of planar structure probe shown in Fig. 3.9, which can be compared to a well-matched dipole when in contact with lossy medium (skin, muscle, water, etc.). This applicator is achieved by means of a gradual transition at constant impedance, from a classical microstrip line to a microstrip line with its ground plane opened. The applicator is formed by the circular ground plane aperture being in contact with

tissue. As this applicator is also used for microwave hyperthermia, one can note in Fig. 3.9 the metallic cylinder placed on the microstrip line in the aperture. This device offers the possibility of central cooling by water flowing in the cylinder, avoiding superficial burning of tissue with microwave energy. The microwave tests we have carried out with this new type of probe show that this applicator presents microwave performances better than the waveguide type, as can be seen on Fig. 3.10 for the reflection coefficients (S_{11} parameter).

3.4 Control of Hyperthermia by Microwave Radiometry

3.4.1 Principles and Problems

Hyperthermia therapy by microwave energy consists in heating tumoral tissues by means of an applicator. This process must take into account the fact that healthy tissues cannot bear temperatures higher than 43 °C; consequently the heating of the volume subjected to microwaves has to be controlled and the power generator monitored. To this purpose, systems combining microwave heating with temperature measurements by implanted thermocouples are classically used, but nowadays the only atraumatic way to achieve such temperature control is to use a radiometric method [14, 15].

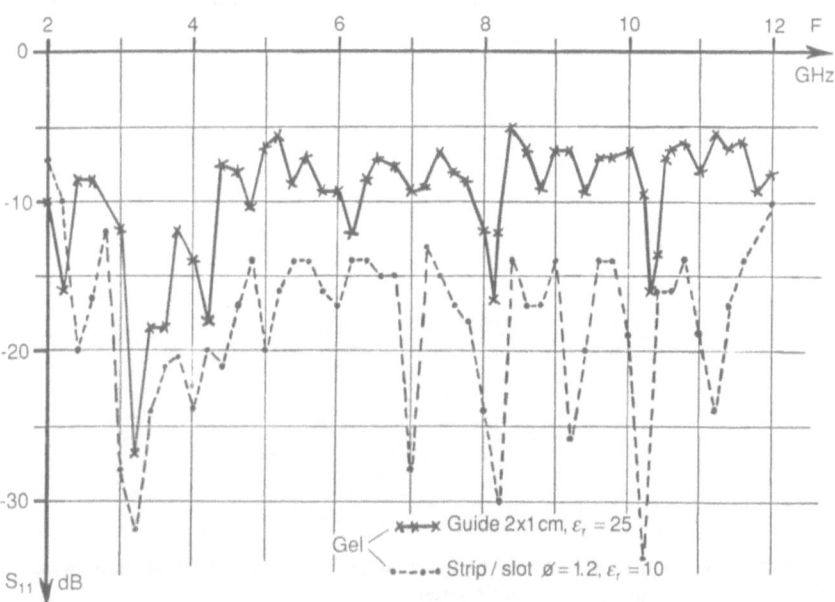

Fig. 3.10. Reflection coefficient S_{11} versus frequency measured on a polyacrylamide gel simulating muscle for two types of applicator (guide and strip-slot)

Gel $\begin{cases} \text{×—×—× Guide 2×1 cm, } \varepsilon_r = 25 \\ \text{•----•----• Strip / slot } \varnothing = 1.2, \varepsilon_r = 10 \end{cases}$

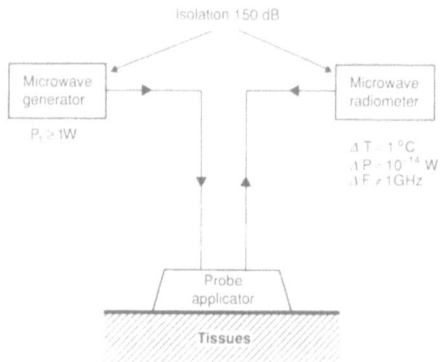

Fig. 3.11. The intermodulation problem in a microwave hyperthermia system combining a generator with a radiometer

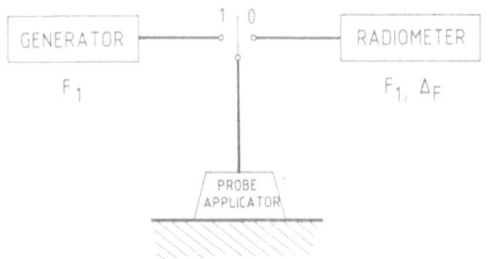

Fig. 3.12. Principle of the alternate method

Such a system as has been developed by the Centre Hyperfréquences et Semiconducteurs is characterized by the use of the same applicator for heating and for radiometric measurement (Fig. 3.11). Its achievement necessitates the avoidance of any intermodulation from the generator to the radiometer. In a typical radiometer the amount of thermal noise power cor-

responding to a temperature variation of 1 °C is nearly 10^{-14} W; moreover the present system is concerned with heating power greater than a few watts so a 150 db minimum value isolation between the two branches (heating and radiometry) of the system is necessary.

In accordance with this principle our system operates according to the alternate method (Fig. 3.12); that is to say, heating is achieved at any time (switch in position 1), radiometric temperature measurements being made over short intervals (for example 10 s every 1 or 2 min) when the switch is in position 0. To avoid any intermodulation from the generator to the radiometer, the heating power is switched off during the radiometric measurements.

3.4.2 Design of the Microwave Systems

Over the last 5 years our microwave systems combining microwave radiometers (operating in the $1-2$ GHz and $2-4$ GHz frequency ranges) with heating generators (at 434 MHz, 915 MHz, or 2450 MHz) have been built and tested and used in different clinics to demonstrate their feasibility. In Lille over the last 3 years more than 2000 heating sessions have been performed on 180 patients, and the results obtained have confirmed the potential, the performances, and the great interest of these heating systems using radiometry to control the temperature [16, 17]. As an example, Fig. 3.13 shows a block diagram of the microwave hyperthermia system HYLCAR, including one generator (60 W at 915 MHz), two radiometers, and a microcomputer to manage the different func-

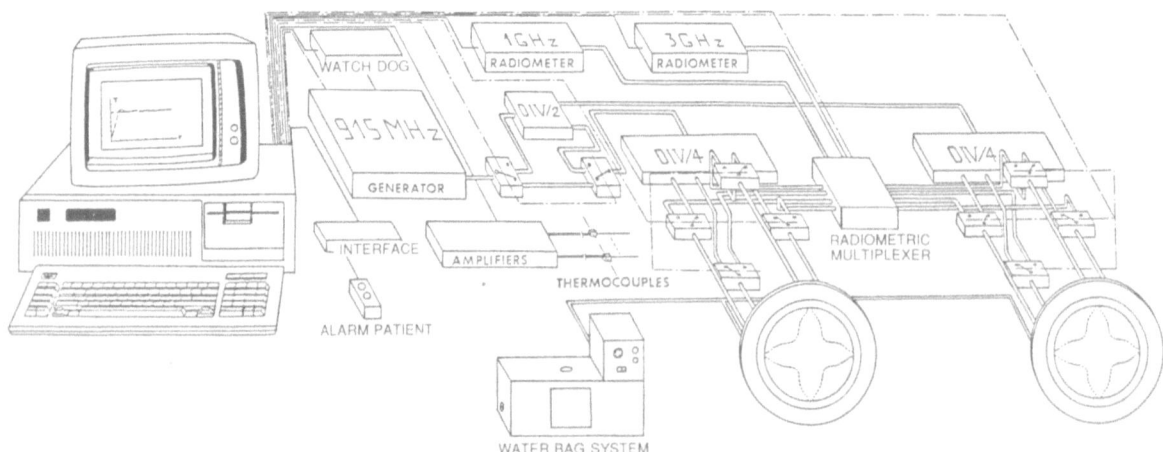

Fig. 3.13. Block diagram of the HYLCAR system which associates a 915-MHz heating generator with two radiometers. Probe applicators are microstrip-microslot multiapplicators

tions of this system. The applicator (Fig. 3.14), which is used for heating and radiometry, is a microstrip-microslot multiapplicator defined from our studies on the single strip-slot applicator. It consists of one circular aperture opened in the ground plane of four microstrip lines with gradual transition, like the first single type applicator. A four-way in-phase power divider feeds the four strip lines, and a central and peripheral cooling system by water flow completes this new applicator.

3.4.3 Performances and Possibilities: Examples of Experiments on Phantoms, Animals, and Patients

As can be noted from Fig. 3.15, this multiapplicator gives a larger and more uniform thermal distribution. The isotherms shown are the result of the heating, for 1 h, of an acrylamide phantom which simulates high water content tissue like muscle; they are deduced from temperature measurements achieved by thermocouples inserted at different depths in the phantom [18, 19]. In this case the radiometric temperatures, measured in the steady state after heating for 1 h, are 42 °C with the 1-GHz radiometer and 43 °C with the 3-GHz one. One can also note the great interest of the central and peripheral cooling system which maintains the superficial temperature at a low value.

Numerous experiments on phantoms and animals have been performed with the HYLCAR system using a single strip-slot applicator or the new multiap-

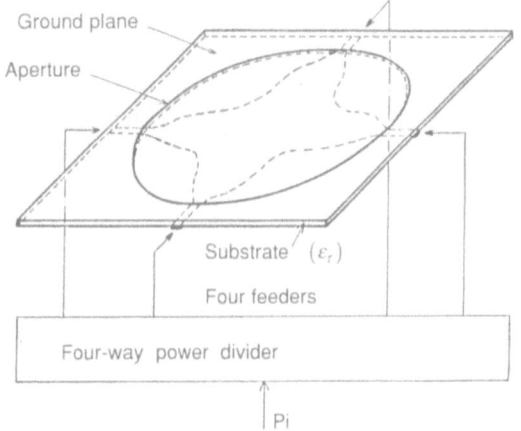

Fig. 3.14. Schematic of the strip-slot multiapplicator

plicator before its use on patients for cancer treatment. Figure 3.16 presents temperature variations versus time recorded during a hyperthermia session on a dog thigh: the radiometric temperature presents the same evolution as the thermocouple temperatures. More than 2000 hyperthermia sessions on 180 patients using systems like HYLCAR have been performed by Dr. Giaux and his team at the Centre Anti-Cancer O. Lambret (Lille, France); these sessions have proved the feasibility, the levels of performances, and the great interest of such systems controlled by radiometry [20]. To illustrate these results, Fig. 3.17 presents the temperature profile on the applicator axis obtained with thermocouples inserted at different depths in tissue and tumor, after 45 min heating of a cervical node [21]. In this case a thermostated water

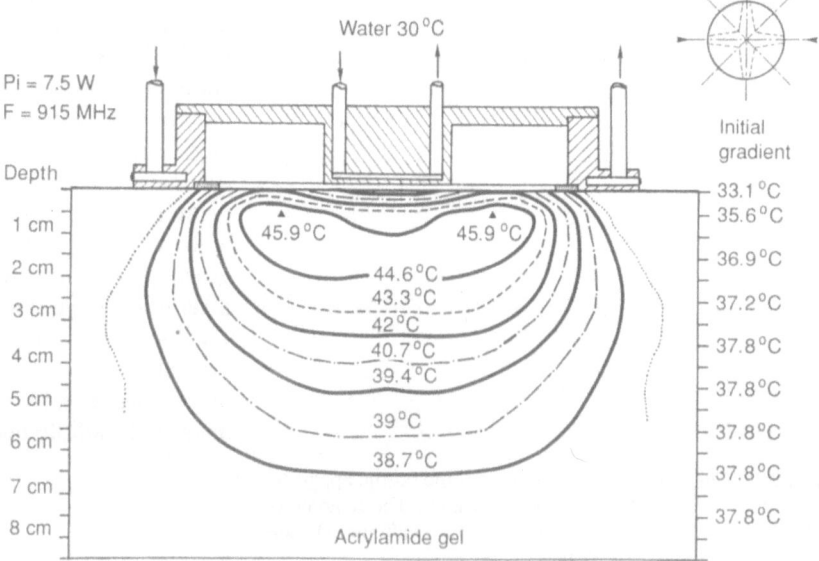

Fig. 3.15. Isotherms in the acrylamide gel after 1 h heating with the multiapplicator (heating frequency 915 MHz; microwave incident power P_i = 7.5 W)

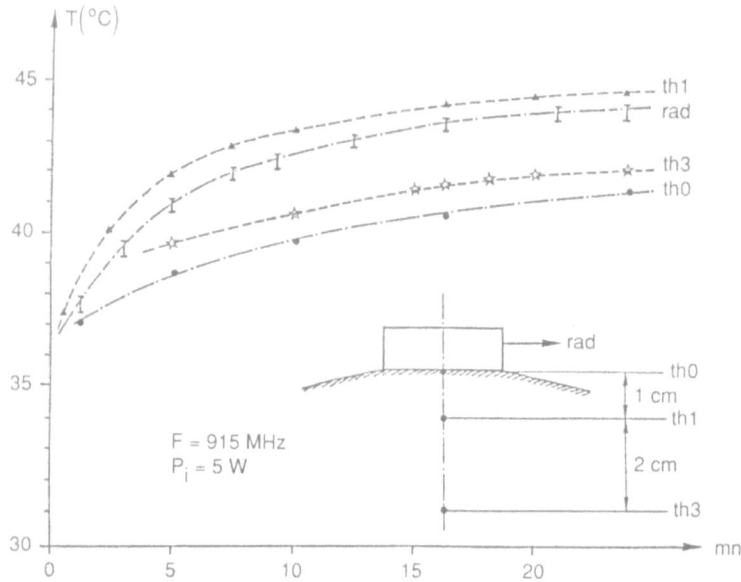

Fig. 3.16. Radiometric (rad) and thermocouple (th0, th1, th3) temperature profiles versus time measured during an hyperthermia session on a dog thigh

bolus placed between the applicator and the skin was used. The measured radiometric temperatures (38.5 °C at 3 GHz and 42.5 °C at 1 GHz) indicate that the temperature bolus contributes to the radiometric signals.

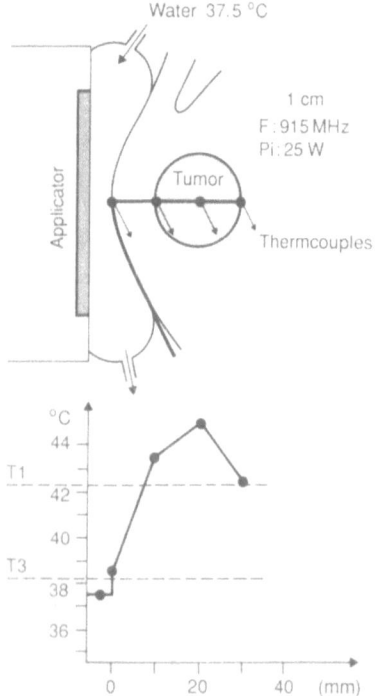

Fig. 3.17. Thermocouple temperature profile on the applicator axis after 45 min heating of a cervical node. The radiometric temperatures are $T_1 = 42.5$ °C (1-GHz radiometer) and $T_3 = 38.5$ °C (3-GHz radiometer)

3.4.4 Use of Microwave Radiometry to Control Radiofrequency Hyperthermia

In Lille, radiofrequency hyperthermia is achieved by means of a classical apparatus based on the well-known capacitive process. The temperature during hyperthermia classically is measured by means of thermocouples inserted in tissue or tumor. As this temperature control is an invasive technique, we have associated with this capacitive system a 3-GHz radiometer as a complementary and atraumatic process. To avoid any intermodulation from the generator to the microwave radiometer, the alternate method has been used: the radiofrequency generator is switched off during the short time devoted to radiometric temperature measurement.

Feasibility experiments have been performed on living anesthetized animals, and comparisons between thermoprobe and radiometric temperature measurements have shown the possibility of substituting the traumatic use of thermocouples by noninvasive temperature control using radiometry. This was confirmed by clinical experiments on patients, and Fig. 3.18 gives the results obtained during a heating session on a cervical lymph node metastasis.

This noninvasive technique for temperature control, which is especially interresting in the case of repetitive heating sessions, is now used in Lille for cancer treatment with radiofrequency hyperthermia [22].

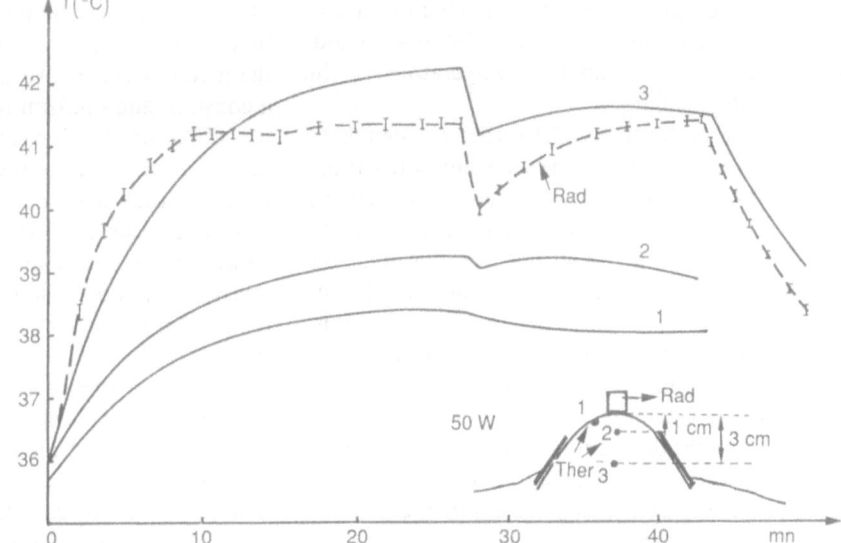

Fig. 3.18. An example of temperature profiles (radiometric: rad, inserted thermocouples: 1, 2, 3) versus time recorded during a radiofrequency capacitive (13.56-MHz) heating session for a cervical lymph node metastasis (the RF power was 50 W)

3.5 Thermal Dosimetry for Microwave Hyperthermia Based on Microwave Radiometry

3.5.1 Principles of the Method

From the results obtained with a microwave hyperthermia system using radiometry for temperature control, noninvasive thermal dosimetry based on the radiometric temperature measurements and the superficial temperature at the applicator-tissue interface during a heating session has been developed [23, 24]. To obtain the thermal profile on the probe axis in the heated medium two numerical programs have been developed and coupled.

The first program calculates the thermal noise power emitted by planar homogeneous or multilayered lossy materials for a TEM propagation, particularly in the case of nonuniform temperature distribution; these calculations give the radiometric temperatures at 1 and 3 GHz.

The second program calculates the one-dimensional thermal profile, based on the bioheat equation, versus depth which occurs when the investigated medium is heated by a defined power. From this calculated gradient the corresponding radiometric temperatures can be deduced by means of the first program and compared with the experimental values. Then using an inverse process by coupling these two programs, taking into account the experimental data, the thermal profile on the applicator axis in the heated medium can be deduced.

3.5.2 Computation of the Radiometric Signal

It is well known [25 – 29] that the radiometric temperature of a planar homogeneous or layered lossy material can be calculated, for a TEM propagation at normal incidence, by the integration of the thermal noise contribution of different path sections dz of the stratified lossy medium under investigation (Fig. 3.19). The power emitted near a path section dz in the medium i is a classical result.

$$2\alpha_i k T(z) \Delta f \, dz \tag{3.23}$$

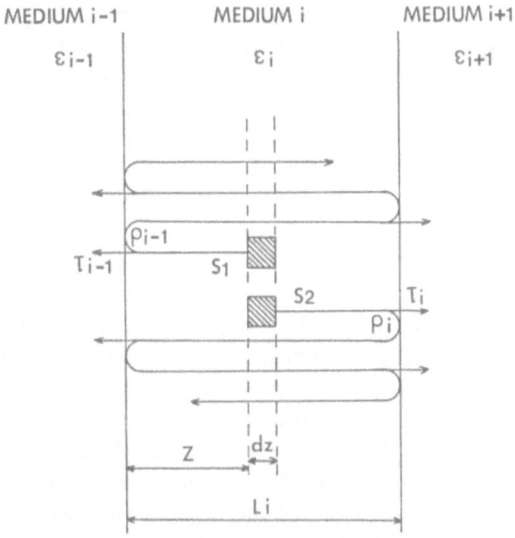

Fig. 3.19. Thermal noise signals arising from the submedium i in the case of a multilayered medium

where α_i = the attenuation of the medium i, k = Boltzmann's constant, Δf = the radiometer bandwidth, and $T(z)$ = the absolute temperature at the depth z in the medium.

As can be noted from Fig. 3.19, the two signals S_1 and S_2 arising from the same subvolume (S_2 propagates backwards and S_1 forwards) are correlated and their contributions are added taking into account the multiple reflections at the interfaces which limit the medium i. This medium is characterized by its thickness L_i, its reflection coefficient Re_i, and its transmission coefficient Tr_i. These multiple reflections introduce phase shifts and attenuation effects, and the signal emerging from the medium i propagates through the different mediums (i-1, i-2, etc.) to the surface. $dP_i(z)$ the contribution to the total thermal signal due to each patch section of the medium i is then expressed by:

$$dP_i(z) = \prod_{j=1}^{j=i-1} |Tr_j|^2$$

$$\times \frac{|\exp(-\gamma_i z) + \varrho_{i+1} \exp[-\gamma_i(2L_i-z)]|^2}{|1+\varrho_{i+1} Re_{i-1} \exp(-2\gamma_i L_i)|^2}$$

$$\times 2\alpha_i k T(z) \Delta f \, dz \qquad (3.24)$$

with

$$Re_i = \varrho_i + \frac{(1-\varrho_i)^2 \cdot \varrho_{i+1} \cdot \exp(-2\gamma_i L_i)}{1+\varrho_i \cdot \varrho_{i+1} \cdot \exp(-2\gamma_i L_i)}, \qquad (3.25)$$

$$Tr_i = \frac{\tau_i \cdot \tau_{i+1} \cdot \exp(-2\gamma_i L_i)}{1+\varrho_i \cdot \varrho_{i+1} \cdot \exp(-2\gamma_i L_i)} \qquad (3.26)$$

where ϱ_i, ϱ_{i+1}, τ_i, τ_{i+1} are respectively the reflection and transmission coefficients as defined in Fig. 3.19 and γ_i is the propagation constant through the medium i. The total emitted noise power for a stratified medium is then obtained with a summation of the emitted signal due to each medium

$$P = P_1 + P_2 + \ldots + P_i + \ldots + P_n . \qquad (3.27)$$

The curvature of the various tissue layers is neglected. At microwave frequencies the layered slab adequately models the flat regions of the human body [30, 31]. A numerical program has been achieved to calculate the noise power emerging from the investigated medium to the applicator; it takes into account the discontinuity at the probe-tissue interface and the following experimental conditions: the central frequency of the radiometer and its bandwidth, the dielectric constants of each medium, and the applicator type used for the radiometric measurement by accounting for the experimental reflection and transmission coefficients at the applicator-skin interface [32].

This program is initially used to compute the theoretical calibration curve of the radiometer when the medium under investigation is supposed to be at a constant and uniform temperature. The calculation for different but increasing temperatures shows a linear variation of the radiometric signal versus temperature; thus these calculations give the theoretical calibration curve of the radiometer for a defined stratified lossy medium. This calibration can also taken into account a water bolus as first medium, the temperature and thickness of which should be constant.

The program then computes the radiometric signal corresponding to the thermal profile versus depth which occurs when the medium in question has been heated. The theoretical radiometric temperature deduced from this calculated signal and the previous calibration curve can then be compared with experimental values.

Comparison between theory and experiments on phantoms has always shown a good correlation whatever the experimental conditions [17, 33]. We have obtained: T_{rad} calculated = T_{rad} experimental $\pm 0.5\,°C$.

For all these measurements we have noticed that each thermal profile in the steady state on the applicator axis is uniquely associated with one "signature", which consists of the surface temperature at the probe-tissue interface, and the two radiometric temperatures (at 1 and 3 GHz). This property, related to the simulation of the thermal profile through the bioheat equation, has allowed us to determine the likely thermal profile on the applicator axis which occurs in the steady state of a hyperthermia session on a lossy medium. Experimental hyperthermia conditions and non-invasive temperature measurements (radiometric and superficial) are used for this determination.

3.5.3 Computation of Thermal Profile Using the Bioheat Transfer Equation

The spatial and temporal distribution of temperature in living tissues is described most commonly by the bioheat transfer equation, which for the one-dimensional multilayered model is:

$$d(z) \cdot c(z) \cdot \frac{\partial}{\partial t} T(z,t) = \frac{\partial}{\partial z}$$

$$\times \left[K_t(z) \cdot \frac{\partial}{\partial z} T(z,t) + B(z,t) + P_a(z,t) + Q_m(z) \right]$$

$$(3.28)$$

where z, t are the space and time variables, $T(z, t)$ the absolute temperature, $d(z)$ the tissue density, $c(z)$ the tissue-specific heat, $K_t(z)$ the tissue thermal conductivity, $P_a(z, t)$ the rate of energy input due to microwave absorption, $Q_m(z)$ the rate of metabolic heat generation, and $B(z, t)$ the rate of heat exchange with blood

The convective heat transfer term $B(z, t)$ is assumed to be equal to the thermal energy brought in by the arterial blood minus the thermal energy carried away with the venous blood.

$$B(z, t) = V_s(z) [T_a - T(z, t)] \qquad (3.29)$$

where T_a is the arterial blood temperature and $V_s(z)$ the product of flow and heat capacity of blood. Note that without the losses involving the blood flow in tissue and the superficial heat exchange, a steady state cannot be reached [34–37].

The biological tissues under investigation are assumed to be heated by a microwave plane wave normally incident. The rate of absorbed power $P_a(z, t)$ is calculated using the active part of the radiative transfer theory [31, 33].

At the boundaries of the medium, the heat flux across the two end planes is assumed to be proportional to the temperature difference with the surrounding media; that is to say:

$$T(\infty, t) = T_a \quad \text{for} \quad t > 0 , \qquad (3.30)$$

$$K_t(0) \cdot \frac{\partial T(0, t)}{\partial z} = h [T(0, t) - T_e] \qquad (3.31)$$

where h is the thermal conductance between the skin and the cooling part of the applicator and T_e the temperature outside the medium.

According to the method of discrete finite differences, the bioheat equation is discretized in space by replacing the differential operator $\partial / \partial z$ by its difference expression. A system of coupled first order differential equations is then obtained, the matrix form of which is:

$$\frac{\partial \bar{T}}{\partial t} = A \bar{T} + \bar{S} \qquad (3.32)$$

where $\bar{T}$ is the temperature vector in sublayers and $\bar{S}$ the source function in the sublayers.

In the steady state $\partial \bar{T} / \partial t = 0$ the matrix A is tridiagonal.

The system of differential equations is solved using the gaussian reduction procedure.

Figure 3.20 gives an example of hyperthermia simulation. The thermal profile in the steady state which exists in the indicated biological multilayered medium irradiated by an incident microwave power density of 0.2 W/cm^2 at 2.45 GHz has been deduced from the computation. The 2.4 GHz radiometric temperature corresponding to this thermal profile calculated with the first program gives $T_{rad} \# 38.5 \,°C$.

3.5.4 The Inverse Process

Using the experimental data imposed (heating frequency, incident microwave power), known (dielectric and thermal characteristics of the different tissues

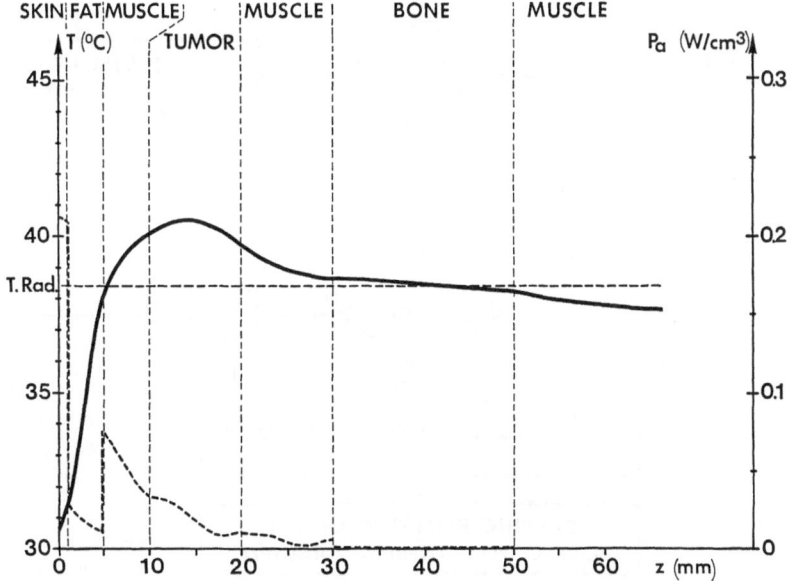

Fig. 3.20. Hyperthermia simulation of a multilayered tissue: temperature profile in the steady state versus depth resulting from irradiation with an incident power density of 0.2 W/cm^2 at 2.45 GHz. The *dotted line* corresponds to the absorbed power variations in each layer versus depth

constituting the heated medium), or measured (superficial and radiometric temperatures, reflection coefficient at the applicator-skin or -bolus interface) during the hyperthermia session, the two previous programs are coupled by optimization loops concerning thermal parameters and radiometric temperatures (at 3 GHz and 1 GHz) as indicated on Fig. 3.21, which shows a block diagram of the computer programs. In this way a simulated temperature curve versus depth in the heated medium is obtained, for which the superficial and the two radiometric temperatures correspond to the experimental values.

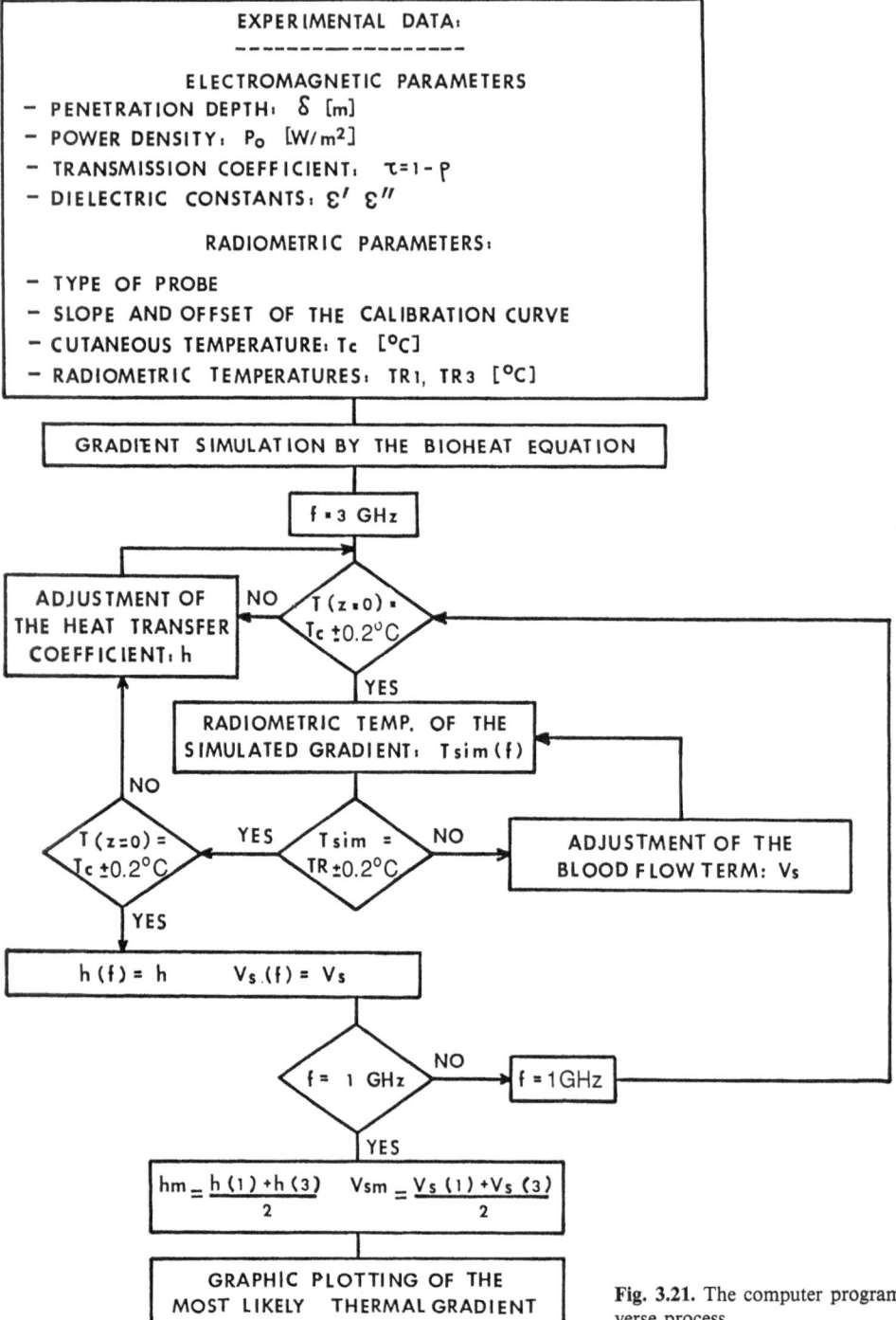

Fig. 3.21. The computer program simulation for the inverse process

3.5.5 Examples of Results Obtained with These Programs: Theory-Experiment Comparison

We have used this inverse process to determine the thermal profile on the applicator axis in the case of hyperthermia first on animals [23], then on patients [38]. In all cases we have obtained a thermal profile very close to the experimental one (deduced from the inserted thermocouple temperature measurements). Figures 3.22 and 3.23 show examples of this comparison in the cases of microwave heating of a dog thigh and hyperthermia treatment of a tumor: an excellent correlation between theory and experiment can be noted.

3.6 Conclusion

Microwave radiometry provides a noninvasive way to measure, control, and monitor the temperature which results from the heating (microwave or radiofrequency) of a biological medium (living tissues+tumors, etc.). This is of great interest in the case of repetitive heating sessions for cancer treatment.

The computerization we have developed for the one-dimensional and multilayered model based on microwave radiometric measurements and hyperthermia simulation constitutes the first step in atraumatic thermal dosimetry in hyperthermia process for tumor treatment. Research will now be aimed at extending this method to a two-dimensional multilayered model.

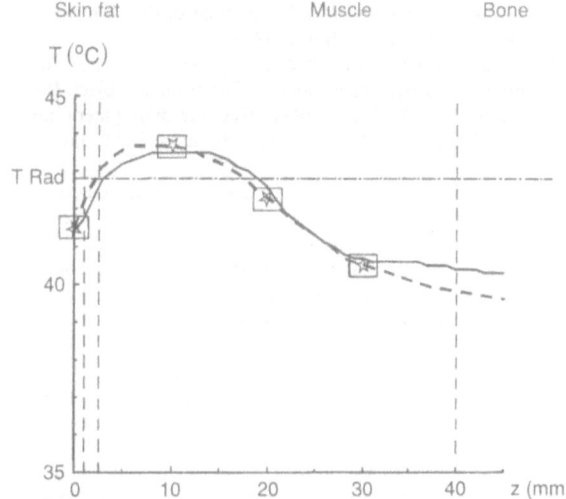

Fig. 3.22. Inverse process: comparison between the measured temperatures (**) and the computed profile on the probe axis after 45 min heating of a dog thigh (P = 7 W; f = 915 MHz) using a strip-slot applicator (ε_r = 4.9; $\oslash$ = 5 cm)

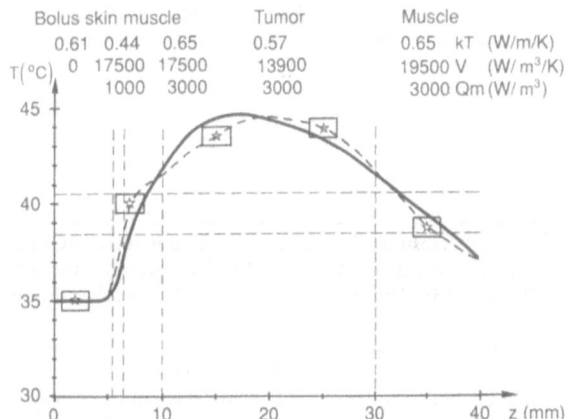

Fig. 3.23. Inverse process: temperature profile computation as compared with inserted thermocouple temperatures measured on the axis of the strip-slot applicator after 1 h heating of cervical node (P_i = 25 W; heating frequency: 915 MHz; radiometric temperatures: $T_{1\,GHz}$ = 40.5 °C, $T_{3\,GHz}$ = 38.5 °C). In this case a thermostated bolus (35 °C) was used

References

1. Dicke RH (1946) The measurement of thermal radiation at microwave frequencies. Rev Sci Instr 17(7):268–275
2. Barret AH, Myers PC (1975) A method of detecting subsurface thermal patterns. Biol Radiol 6:45–56
3. Mamouni A, Bliot F, Leroy Y, Moschetto Y (1977) A modified radiometer for temperature and microwave properties measurements of biological substances. In: Proceeding of the 7th Eur Microw Conf. Copenhagen, p 703
4. Schwan HP, Foster KR (1977) Microwave dielectric properties of tissue. Biophys J 17:193–197
5. Schepps JL, Foster KR (1980) The UHF and microwave dielectric properties of normal and tumor tissues: variation in dielectric properties with tissue water content. Phys Med Biol 26(6):1149–1153
6. Chive M, Constant E, Leroy Y, Mamouni A, Moschetto Y,

Nguyen DD, Sozanski JP (1981) Procédé et dispositif de thermographie-hyperthermie en microondes. Brevet d'invention no 81 00682, Jan 1 1981.
7. Chive M et al. (1982) Method and apparatus for measuring temperature of body. Japan Patent no 57 001394, Jan 9 1982.
8. Chive M et al. (1982) Process and apparatus for measuring temperature of a body by microwaves. U.S. Patent no 342 464. Jan 25 1982
9. Nguyen DD (1980) Thermographie et chauffage microonde: contribution à la conception et à la réalisation de systèmes destinés au génie biologique et médical. Thèse de 3ème cycle, 1980, Lille
10. Chive M, Plancot M, Vandevelde JC (1983) Wideband microstrip-slot applicators for microwave hyperthermia and

microwave thermography. In: Proceedings of the 18th IMPI symposium, July 1983. Philadelphia

11. Ringeisen, V., Chive M, Toutain S (1983) Applicator for supplying radiofrequency energy to and from an object. US patent no 569400, Jan 9 1984, West German patent no P 33006776, Jan 11 1983, French patent no 8400363, Jan 11 1984

12. Ledee R, Chive M, Plancot M (1984) Antennes plaquées pour applications biomédicales. In: Proceedings of the IV journées nationales microondes. June 1984. Lannion

13. Ledee R, Chive M, Plancot M (1985) Microstrip microslot antennas for biomedical applications: frequency analysis of different parameters of this type of structure. Electronics Letters, March 28 1985, vol 21, no 7

14. Nguyen DD, Mamouni A, Leroy Y, Constant E (1979) Simulations microwave local heating and microwave thermography. Possible clinical applications. J Microwave Power 14(2):135−137

15. Nguyen DD, Chive M, Leroy Y, Constant E (1980) Combination of local heating and radiometry by microwaves. IEEE Trans Ins Meas IM-29 2:143−144

16. Chive M, Plancot M, Leroy Y, Giaux G, Prevost B (1982) Microwave (1 and 2.45 GHz) and radiofrequency (13.5 GMHz) hyperthermia monitored by microwave thermography. In: 12th European microwave conference proceedings, Sept 1982. Helsinki

17. Plancot M (1983) Contribution à l'étude théorique expérimentale et clinique de l'hyperthermie microonde contrôlée par radiométrie microonde. Thèse de 3ème cycle, Dec 1983. Université des Sciences et Techniques, Lille

18. Chive M, Plancot M, Giaux G, Prevost B (1984) Microwave hyperthermia controlled by microwave radiometry: technical aspects and first clinical results. J Microwave Power 19(4):233−241

19. Ledee R, Playez E, Chive M, Plancot M. Moschetto Y (1986) Microstrip microslot multiapplicator for microwave hyperthermia and radiometry. In: Proceeding of the 8th IEEE-EMBS Conference Nov 1986, vol 3. Forth Worth, pp 1529−1532

20. Giaux G, Delannoy J, Prevost B, Chive M, Plancot M, Delvalee D, Ledee R (1987) Hyperthermie oncologique par microondes contrôlée par radiométrie microonde associée à la radiothérapie: considérations techniques et résultats cliniques. Proc 1er séminaire ODAM: 1'hyperhtermie en cancérologie, 11−13 Mai 1987. Wissembourg

21. Delannoy J (1987) Contribution à l'étude de l'hyperthermie clinique microonde. Application à la réalisation d'un système interactif de traitement par hyperthermie. Thèse de Doctorat, Université Lille II, Lille

22. Giaux G, Prevost B, Delannoy J, Delvallee D (1987) Hpyerthermie oncologique radiofréquences associée à la radiothérapie: considérations techniques et résultats cliniques. In: Proc 1er Séminaire ODAM: l'hyperthermie en cancrèrologie, 11−13 Mai 1987. Wissembourg

23. Plancot M, Chive M, Giaux G, Prevost B (1984) Thermal dosimetry in microwave hyperthermia process based on radiometric temperature measurements: principles and feasibility. Hyperthermic Oncol 1:863−866

24. Plancot M, Prevost B, Chive M et al. (1987) A new method for thermal dosimetry in microwave hyperthermia using microwave radiometry for temperature control. Int J Hyperthermia 3(1):9−19

25. Rain Water JH (1978) Radiometers: electronic eyes that see noise. J Microwave, Sept 1978, 58−62

26. Tsang L, Njonu E, Kong JA (1975) Microwave thermal emission from a stratified medium with non uniform temperature distribution. J Appl Phys 46:5127−5133

27. Wilheitt TT (1978) Radiative transfer in a plane stratified dielectric. IEEE Trans Geo Sci Electron 16:138−143

28. Fabre JJ, Leroy Y (1981) Thermal noise emission of a lossy material for a TEM propagation. Electron Lett 17 (11):376−377

29. Bardati F, Solimini D (1984) On the emissivity of layered materials. IEEE Trans Geo Sci Electron 31:317−322

30. Bardati F et al. (1985) Retrieval of hyperthermia induced temperature distribution from noisy microwave data. Electron Lett 21:800−801

31. Wagter de C (1985) Computer simulation for local temperature control during microwave induced hyperthermia. J Microwave Power 20:31−42

32. Fabre JJ (1982) Méthode de calcul de signaux thermiques et possibilités de nouvelles utilisations de la thermographie microonde. Thèse de 3ème cycle. Université des Sciences et Techniques, Lille

33. Plancot M, Prevost B, Fabre JJ, Chive M, Moschetto Y, Giaux G (1986) Thermal dosimetry based on radiometry in multilayered media. In: Proceedings of the 8th IEEE-EMBS conference, Nov 1986, vol 3. Forth Worth, pp 1429−1431

34. Carlsaw HS, Jaeger JC (1959) Conduction of heat in solids. Oxford University Press, Oxford

35. Cravalho EG, Fox LR, Kan JC (1980) The applicator of the bioheat equation to the design of thermal protocols for local hyperthermia. Ann NY Acad Sci 335:86−97

36. Roemer RB, Cetas TC (1984) Application of bioheat transfer simulations in hyperthermia. Cancer Res 44:4788−4798

37. Wulff W (1974) The energy conservation equation for living tissue. IEEE Trans Biomed Eng 21:494−496

38. Plancot M (1987) Dosimétrie thermique en hyperthermie microonde basée sur la radiométrie microonde et l'équation de la chaleur. In: Proc 1er séminaire ODAM: l'hyperthermie en cancérologie, 11−13 Mai 1987. Wissembourg

Subject Index

Clinical Thermology

Subseries Thermotherapy

Series Editor: M. Gautherie

Biological Basis of Therapeutic Hyperthermia

Methods of External Heating

Methods of Hyperthermia Control

Thermal Dosimetry and Treatment Planning

Whole-Body Hyperthermia in Clinical Oncology

Interstitial, Endocavitary and Perfusional Hyperthermia –
Methods and Clinical Trials

Subseries Diagnosis

Series Editor: M. Gautherie

Thermal Measuring, Recording and Imaging Techniques

Chronobiology and Thermophysiology

Clinical Thermology of Breast Diseases

Clinical Thermology of Rheumatic and Bone Diseases